Solar PV Panels - Recent Advances and Future Prospects

Edited by Basel I. Ismail

Published in London, United Kingdom

Solar PV Panels - Recent Advances and Future Prospects
http://dx.doi.org/10.5772/intechopen.104056
Edited by Basel I. Ismail

Contributors
Fikadu Takele Geldasa, Hadi Ahmadi Moghaddam, Matthew Phillips, Svetlana Tkachenko, Victoria Timchenko, Shiva Hayati Raad, Zahra Atlasbaf, Tulshi Shiyani, Abdellah Asbayou, Lahoussine Bouhouch, Ismail Isknan, Ahmed Ihlal, Quentin Lagarde, Bruno Beillard, Serge Mazen, Julien Leylavergne, Ramesh Sharma, Debidatta Behera, Sanat Mukherjee, Aparna Dixit, Arti Saxena, Abdul Shakoor, Ghazi Aman Nowsherwan, Muhammad Fasih Aamir, Ahmar Ali, Sami Ur Rehman, Waheed Alam, Muhammad Yasir, Khizra Arif, Muhammad Ahmad, Jamal Yousaf, Khalil Kassmi, Olivier Deblecker, Ali Lamkaddem, Rachid Malek, Najib Bachiri, Hajar Chadli, Khalid Salmi

First published in London, United Kingdom, 2023 by IntechOpen
IntechOpen is the global imprint of INTECHOPEN LIMITED, registered in England and Wales, registration number: 11086078, 5 Princes Gate Court, London, SW7 2QJ, United Kingdom

British Library Cataloguing-in-Publication Data
A catalogue record for this book is available from the British Library

Additional hard and PDF copies can be obtained from orders@intechopen.com

Solar PV Panels - Recent Advances and Future Prospects
Edited by Basel I. Ismail
p. cm.
Print ISBN 978-1-83768-676-6
Online ISBN 978-1-83768-677-3
eBook (PDF) ISBN 978-1-83768-678-0

Meet the editor

Professor Dr. Basel I. Ismail is a licensed professional engineer (PEng) in Canada and an associate professor and the chair of the Mechanical Engineering Department, Lakehead University, Ontario, Canada. He is a pioneering academic researcher in alternative energy and renewable and clean energy engineering technologies.

Contents

Preface **IX**

Chapter 1 **1**
Solar Photovoltaic Principles
by Aparna Dixit, Arti Saxena, Ramesh Sharma, Debidatta Behera and Sanat Mukherjee

Chapter 2 **17**
Transparent Solar PV Panels
by Tulshi Shiyani

Chapter 3 **27**
Photovoltaic and Photothermal Solar Cell Design Principles: Efficiency/Bandwidth Enhancement and Material Selection
by Shiva Hayati Raad and Zahra Atlasbaf

Chapter 4 **43**
Operation of Photovoltaic Panels in Stand-alone Applications
by Ali Lamkaddem, Hajar Chadli, Khalid Salmi, Rachid Malek, Olivier Deblecker, Khalil Kassmi and Najib Bachiri

Chapter 5 **55**
Solar Energy Conversion Efficiency, Growth Mechanism and Design of III–V Nanowire-Based Solar Cells: Review
by Fikadu Takele Geldasa

Chapter 6 **91**
What is the Profitability of a Photovoltaic Installation in France for an Individual?
by Quentin Lagarde, Bruno Beillard, Serge Mazen and Julien Leylavergne

Chapter 7 **119**
Utilization of MOSFET Transistor to Characterize PV Panels under Dust: Study Area Agadir-Morocco
by Abdellah Asbayou, Lahoussine Bouhouch, Ismail Isknan and Ahmed Ihlal

Chapter 8 **135**
Influence of Wind Incidence Angle on the Cooling of Rooftop-Mounted Solar Panels
by Hadi Ahmadi Moghaddam, Matthew Phillips, Svetlana Tkachenko and Victoria Timchenko

Chapter 9 **145**
Performance Evaluation of Solar Cells by Different Simulating Softwares
by Abdul Shakoor, Ghazi Aman Nowsherwan, Muhammad Fasih Aamir, Ahmar Ali, Sami Ur Rehman, Waheed Alam, Muhammad Yasir, Khizra Arif, Muhammad Ahmad and Jamal Yousaf

Preface

Recent environmental concerns about greenhouse gas (GHG) emissions that cause global warming and climate change, as well as the depletion of energy resources, have sparked substantial research into new methods of producing electricity.

Solar energy is one of the most significant renewable energy sources and has been attracting increased attention in recent years due to its abundance as well as its being a pollutant-free and environmentally friendly technology. Solar photovoltaic (PV) technology provides direct conversion of solar energy (photons) into electrical DC power. *Solar PV Panels – Recent Advances and Future Prospects* is the result of contributions from several leading researchers and experts worldwide. This book is an excellent source of information for researchers, academicians, industry experts, energy policymakers, and economists working in renewable energy, specifically solar PV panels for direct electric power production. This book covers topics such as solar PV fundamentals; solar panels modeling, simulations, optimization, and economics; solar panels testing; semiconductor materials for solar cells and panels; solar PV applications and policies; and solar PV technology and battery storage sizing for standalone and grid-connected applications. This book expands scientific and technological knowledge for future renewable solar energy projects.

I would like to sincerely thank all chapter authors for their efforts and the quality of the chapters presented. Also, I would like to thank Author Service Manager Ms. Dajana Jusic at IntechOpen for her excellent efforts in managing the publication process of this book.

Dr. Basel I. Ismail, P.Eng.
Associate Professors and Chair,
Department of Mechanical Engineering,
Lakehead University,
Thunder Bay, Ontario, Canada

Chapter 1

Solar Photovoltaic Principles

Aparna Dixit, Arti Saxena, Ramesh Sharma, Debidatta Behera and Sanat Mukherjee

Abstract

Due to the limited supply of fossil fuels in the modern era, humankind's need for new energy sources is of utmost importance. Consequently, solar energy is essential to society. Solar energy is an endless and pure source of energy. Solar energy research is being used to help solve the world's energy dilemma, safeguard the environment, and promote significant sustainable economic growth. Humans have now constructed numerous solar photovoltaic power plants to produce electricity, and many people have installed solar panels on their homes' roofs to do the same. The non-mathematical explanation of PV solar cell theory and its circuit architecture is covered in this chapter. It is written for a variety of groups, including engineers who need an introduction to the subject of photovoltaic cells, end users who require a deeper understanding of the theory to support their applications, students interested in PV science and technology, and others. The fundamentals of the individual electricity-producing solar cell—the photovoltaic cell—are discussed in this chapter. The reader is informed about the workings of PV cells. The chapter focuses on the operation and construction of PV cells. The advantages and disadvantages of the cell's potential industrial applications are discussed. Here, we go over how to ensure that the PV cells used in contemporary renewable energy systems are up to snuff.

Keywords: new energy, fossil fuels, solar power, solar photovoltaic, inexhaustible energy, clean energy

1. Introduction

When something shifts in the cosmos, it also shifts in a quantity we call energy. Although its exact definition is unknown, the term "energy" is frequently used to ability of a physical or biological system to describe the motion or change. Movement, heating, and chemical transformation are just some of the various changes that can be brought about by applying different types of energy. Energy is required for all actions, including those performed by humans. It's necessary for humans to do activities like walk, cook, heat, and light their homes, and operate motor vehicles. Mankind is a voracious energy consumer. On average, a young man's daily energy needs amount to roughly 2500 kcal (2.9 kWh) while he is actively engaged in physical activity. An annual energy cost of roughly $106 is implied. The present annual worldwide energy use is close to 20,000 kWh per person. If this trend continues, men will consume 19 times more energy than they need to maintain their current level of health and well-being. Over the past 100 years, humanity has seen a significant growth in energy use. While the annual energy consumption per resident was approximately 5800 kWh in 1890, it increased to 20,200 kWh in 1970. Since 1970, the amount of energy used per resident per year has decreased to the current level of 19,000 kWh. The evolution process that began roughly five centuries ago can be linked to the rise in energy use in the twentieth century. The eighteenth century's Enlightenment saw the development of the philosophy of human progress, which provided the framework for this procedure. The method was intended to examine the environment and how it could be modified to better meet the needs of those whose lives would become safer and more comfortable. Growing industrialisation and mass production, which required an increasing amount of energy, went hand in hand with this process. Coal was the primary energy source at the end of the nineteenth century. During this time, electricity—a brand-new and sophisticated kind of energy—was introduced in the industrialised nations. This kind of energy was quickly used extensively. In the early twentieth century, as a result of the enormous rise in electricity demand and the subsequent building of hydroelectric plants, hydropower became a significant source of energy. At the close of the twentieth century, the oil and gas industries began to play a considerable part as contributors to the world's energy supply. Present-day energy sources are dominated by coal, oil, and gas. These three forms of energy (coal, oil, and gas), generally known as fossil fuels, are referred to as "conventional energy sources." During this time, nuclear energy was introduced as a brand-new energy source. Numerous household items are inexpensive because of growing and more efficient mass production. Given the tremendous increase in product consumption, it's not surprising that we now refer to modern society as a "consuming society." The quest for and use of new, sustainable energy sources has become a major issue near the end of the twentieth century. The urgency of this problem is highlighted by the Earth's finite supply of fossil fuels and rising global demand for energy generation. The focus is shifting to renewable energy sources because of this. Every human action requires energy. When considering the production and consumption of energy by human civilization, an energy system can be used to explain it. The population, overall energy consumption, and the types and sources of energy that people consume make up the key elements of the energy system. The energy system at the beginning of the twenty-first century is characterised by the six billion people who occupy the globe and the 1,300,000,000 kW of total energy use.

Even now, the cost of producing electricity from renewable sources is higher than that of conventional ones. As a result, it is not yet economically feasible in industrialised nations to use renewable energy sources as electricity power sources on a large scale. The main driver of a developing market for renewable energy sources and favourable national regulations is certainly their positive effects on the environment. But for the two billion people who, in many parts of the world, do not have access to an electricity grid, power generated from renewable energy sources is already the option that has the lowest overall cost.

The definition of "energy transition" is the move away from carbon-based energy sources during the coming years [1]. Therefore, it is crucial to provide energy using sources that can be regenerated organically over time on a human timeline. Renewable energy is the name given to this type of energy, which is largely environmentally safe and sustainable. Recent technological advancements have made it easy to convert various types of renewable energy into other forms of energy (such as electricity and heat).

As a direct consequence of this, in 2015 the international community came together to adopt the Sustainable Development Goals (SDGs) as a component of the

United Nations 2030 Agenda for Sustainable Development [2]. The objectives promise to end malnutrition, poverty, etc. Some of the objectives specified by the world community, such as the provision of clean energy, climate protection, and others, were energy-related. The seventh goal (SDG-7) aims to raise the proportion of renewable energy in total energy by the year 2030 and to offer clean, cheap, and contemporary energy services in all parts of the world. In addition, Sustainable Development Goals—13 aims to enhance immunity either through increasing the resistance of a number of countries or by educating and raising awareness among the general public.

This is due to the fact that every nation on earth is susceptible to the repercussions of climate change. The vast majority of the energy that is required to keep things moving in the atmosphere comes from the sun. This energy is emitted in the form of electromagnetic waves, which have wavelengths ranging from 0.2 to 4 m and are invisible to the human eye. A photon is the smallest quantifiable unit of an electromagnetic field. The current definition of a photon is based on studies conducted by Albert Einstein between 1900 and 1920 (which were based on studies conducted by the German scientist Max Planck). Gilbert Lewis is credited with being the first person to use the term "photon," which was first published in Nature in 1926. Solar irradiance is the amount of energy that is received from the sun at a certain location and time in the form of electromagnetic radiation. This energy is measured in watts. It is stated as a number, and the unit of measurement used to determine it is W/m^2. One of the most important forms of renewable energy is the sun's rays, which are collectively referred to as "solar energy" by the general public. There are two primary categories of technology that exist to capture the sun's rays: active solar technologies, such as photovoltaic systems, and passive solar technologies, which encompass a wide range of practises and may include orienting a building to the sun. Active solar technologies include photovoltaic systems. Passive solar technologies include a wide range of practises. Active solar technologies include photovoltaic systems.

2. Basic operational principles

Direct use of solar energy can be performed in essentially two different ways: (1) the transformation of sunlight directly into electricity in semiconducting devices that are more popularly known as solar cells; and (2) the collection of heat in solar collectors. The transformation of solar radiation into electrical current is referred to as "photovoltaic energy conversion" (PV), and this is the meaning of the word "photovoltaic energy conversion."

The photovoltaic effect is responsible for this phenomenon. The phenomena that can cause a potential difference to occur at the interface of two materials that are not identical is referred to as the "photovoltaic effect," and it is described using the word "photovoltaic effect." As a result, the entire field that studies the conversion of solar energy into electricity is referred to as "photovoltaics," and its acronym stands for "photovoltaic electrics." The term "photovoltaics" comes from the combination of the Greek word for light ("photo") with the Italian name of an early electrical researcher, Alessandro Volta (1745–1827), which is shortened to "Volt." Since most people do not know what the word "photovoltaics" means, the term "solar electricity" has become the most frequent way of referring to PV solar energy.

Three primary processes are necessary for the photovoltaic effect to take place: (1) the generation of charge carriers as a result of photon absorption by the materials comprising a junction; (2) the subsequent separation of the photo-generated charge carriers within the junction; and (3) the collection of the photo-generated charge carriers at the terminals of the junction.

2.1 History of photovoltaic effect

The discovery of the photovoltaic phenomenon is attributed to a French physicist named Alexandre Edmond Becquerel in the year 1839. During the course of his experiments using metal electrodes and electrolyte, he noticed that the conductivity increased as the amount of light increased. Willoughby Smith made the discovery that selenium possesses photovoltaic properties in the year 1873. Actually, when a material absorbs light at a frequency over a threshold frequency that varies with the substance, electrons are released. This phenomenon is called the photoelectric effect, and it is closely related to the photovoltaic effect. Taking into account the fact that light is presumed to be made up of individual energy quanta (photons), Albert Einstein was able to explain this phenomenon in 1905. This type of photon's energy can be calculated as $E = h\nu$, where h is Planck's constant and ν is the frequency of the light.

In 1921, Albert Einstein published the paper on the photoelectric effect for which he received his sole Nobel Prize. After another decade, the first pure semiconductor was developed in 1931. Solar cells were first used for space applications in the 1950s. In a short amount of time, Hoffman Electronics was able to surpass the previous record for solar cell efficiency, achieving highs of 10 and 14% in the years 1959 and 1960, respectively. The earliest solar cells, which had an efficiency of roughly 8%, were invented in 1957. A short time after that, the first PV cell made of amorphous silicon was developed, and the capacity of PV systems reached 500 kW. This amount continued to rise, reaching its highest point of 21.3 MW in the year 1983. A high-concentrating photovoltaic (PV) facility with a capacity of 175 kilowatts (kW) was finally constructed in the state of Arizona in the United States in the year 2002. A new PV technology efficiency record of 40% was set 4 years later. In 2012, when the global PV capacity hit 100 GW, production prices dropped drastically to $1.25 per watt. The first solar-powered aircraft completed a global flight in 2016 [3]. The global PV power potential is shown in **Figure 1**. Solar photovoltaic generation will increase by 23 percent, from 156 GWh in 2015 to 821 GWh in 2020, making it the fastest-growing renewable energy source after wind and ahead of hydropower. PV capacity additions experienced an exceptional rise (a record of 134 GW) in China, the US, and Vietnam. Unquestionably, solar PV is moving toward becoming the most affordable choice for producing power globally, and in the years to come, it is anticipated to draw significant investment [4].

2.1.1 Construction and working of photovoltaic cell

Photovoltaic cells are a type of electrical device that are capable of transforming the energy from light into electric current. The solar cell is an example of a photovoltaic cell. This type of cell is often referred to as a PV cell, which is an abbreviation for "photovoltaic cell." A solar cell is composed of its most fundamental component, a diode with a p-n junction. Photoelectric cells, of which solar cells are a type, are devices in which the presence of light causes a change in the electrical properties of the device (such as the current, the voltage, or the resistance). One sort of cell that fits within this category is the solar cell.

In the process of fabricating solar panels, modules that have been constructed from solar cells and then joined together are utilised. These modules are then

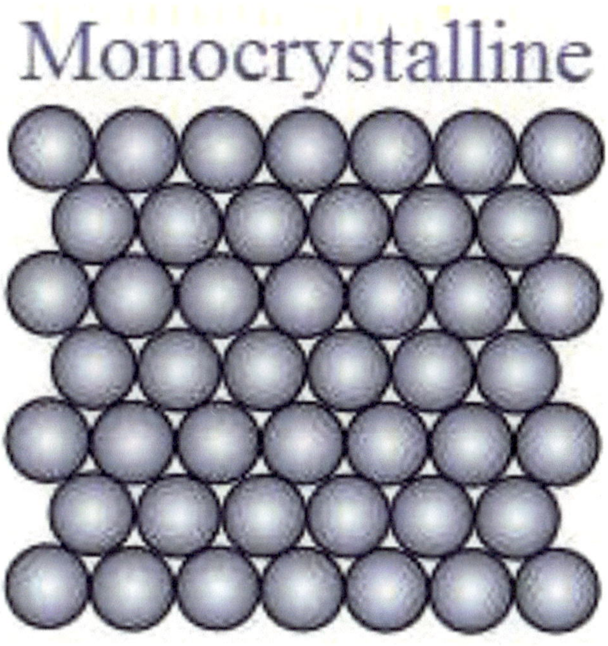

Figure 1.
Monocrystalline structure.

incorporated into panels. When producing at their best levels of efficiency, the vast majority of single-junction silicon solar cells have an open-circuit voltage that falls somewhere in the range of 0.5 to 0.6 volts on average. This occurs when the cells are functioning at their optimal levels. This by itself is not a particularly large quantity, but you need to keep in mind that these solar cells are on the diminutive side. An enormous solar panel has the potential to provide a considerable amount of energy that is kind to the environment.

2.1.1.1 Construction of a solar cell

Although solar cells differ slightly in manufacture from standard p-n junction diodes, they are essentially the same thing. On top of the far more substantial n-type semiconductor, which serves as the base for the formation of the p-type semiconductor, an extremely thin layer is added (**Figure 2**). On top of the p-type semiconductor layer, we then place a few thinner electrodes.

When these electrodes are utilised, the p-type layer does not have its accessibility to light inhibited in any way. A p-n junction can be found just beneath the p-type layer in this structure. The underside of the n-type layer is likewise equipped with an electrode for collecting current. So that the solar cell is not damaged by accidental mechanical disturbance, we enclose the whole thing in a thin glass enclosure.

2.1.1.2 Working of a solar cell

Because the p-type layer is so thin, light photons can easily access the p-n junction. This allows light to go through the material more quickly. A sufficient quantity of energy is supplied to the circuit by the photons of light, which enables the circuit to

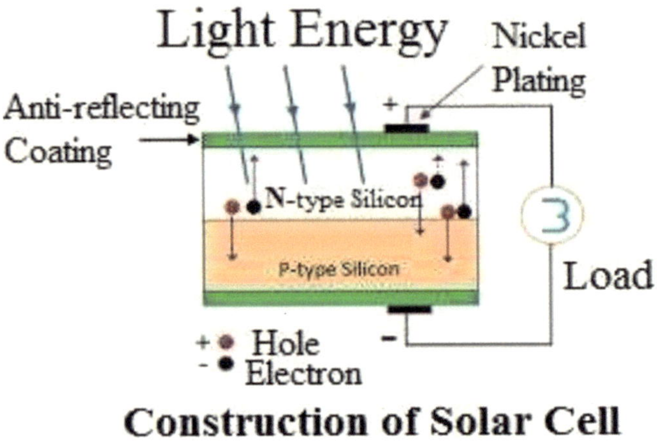

Figure 2.
Construction of solar cell.

generate a significant number of electron-hole pairs. Because of the light, the link is no longer in a condition of thermal equilibrium. This is because of the light. Free electrons have an easier time making their way to the n-type side of the junction when they are located in an area that is deficient in electrons. The depletion produces holes that have the capability of rushing to the P-type side of the link, where they can cause further damage. The barrier potential of the junction halts the freshly formed free electrons in their tracks as soon as they reach the n-type side, rendering them unable to continue crossing the junction. This occurs as soon as the electrons reach the n-type side.

Once they have reached the p-type side of the junction, newly generated holes are unable to continue crossing the junction because of the constant barrier potential that exists between the two sides of the junction. When the number of electrons on one side of a p-n junction (the n-type side) is greater than the number of holes on the opposite side of the junction, the p-n junction behaves like a small battery cell. This occurs when the number of electrons on the n-type side of the junction is greater than the number of holes on the other side of the junction (the p-type side). The voltage that is generated is known as "photovoltage," which is only another word for what it is. When there is only a light load placed across the junction, there will only be a very small amount of current flowing through it.

2.1.2 Basic structure of a photovoltaic cell

An anti-reflective coating (ARC), which is found in the vast majority of existing technologies for solar cells [5, 6], is an anti-reflective coating that is essential to the fabrication of solar cells. Because silicon has such a significant surface reflection, it is often sprayed over cells that are made entirely of silicon.

- The front contacts that is necessary to gather the current from the solar cell (**Figure 3**). They frequently construct with metal materials.

  An emitter that absorbs photons and then transfers the energy from those photons to an excited state that carries a charge after doing so. Since pentavalent-doped (n-type) silicon has a higher surface quality than trivalent-doped

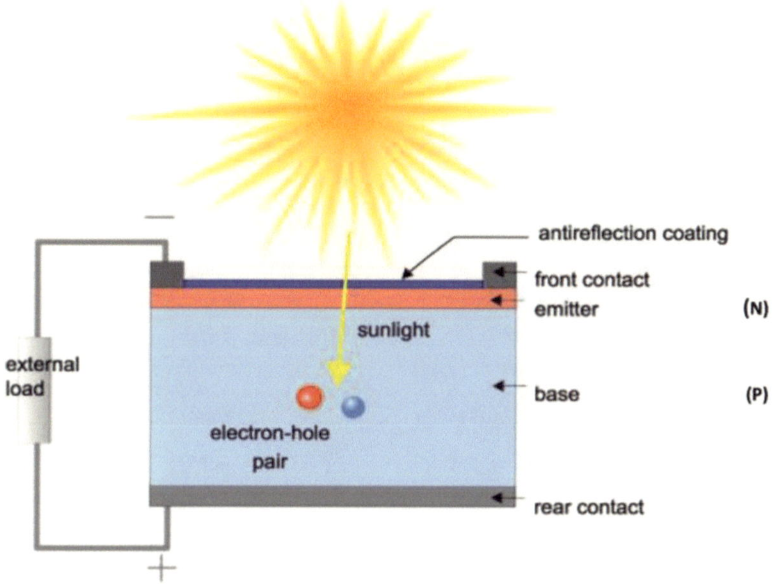

Figure 3.
Basic structure of photovoltaic cell.

(p-type) silicon, it is placed in the front of the cell, which is where the majority of the light absorption takes place. This is because pentavalent-doped (n-type) silicon is more expensive than trivalent-doped (p-type) silicon.

- In a p-n junction, the majority of the electrons in the region of the n-type side of the junction that is closest to the junction diffuse to the p-type side, and the majority of the holes in the region of the p-type side that is closest to the junction diffuse to the n-type side.

- The front contact is more critical than the rear contact since it is closer to the junction; nevertheless, the rear contact does not need to be transparent because it is further away from the junction.

2.1.3 Different types of PV cell

Typically, there are four primary kinds of solar cells, which include the following: (1) Monocrystalline solar cells, (2) polycrystalline cells (multi-silicon cells), (3) amorphous solar cells, and (4) thin film solar cells.

(1) *Monocrystalline solar cells*, which are also known as single crystalline cells (**Figure 1**), can be differentiated from other types of solar cells by their dark black appearance. As a result of this look, monocrystalline solar cells are easily identifiable in comparison to other forms of solar cells. They are made using an incredibly pure type of silicon, which is what has driven them to the position of being the most effective material for the process of converting sunlight into electricity. As a result of this, solar panels have risen to the position of being the most effective material. Solar photovoltaic cells made from monocrystalline silicon have an efficiency of 15%*, making them the most effective but also the most expensive option. They are capable of producing up to four times as much electricity as

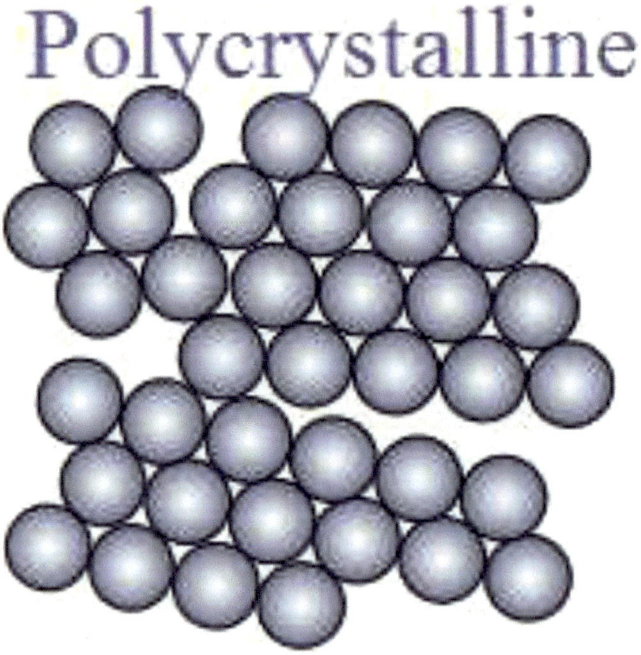

Figure 4.
Polycrystalline.

thin-film solar panels, but they only take up a small fraction of the space. In environments with low light, their performance is superior to that of other panels, and they have a longer lifespan. The most major downside of this option, as well as the primary reason why it does not enjoy more popularity among homeowners, is the high cost. In addition, dust and shadows can cause the circuit to malfunction, and the method of manufacturing that involves cutting the cells into wafers is usually regarded as being inefficient.

(2) At the very beginning of the 1980s, the solar industry developed polycrystalline cells (**Figure 4**), which are also known as *multi-silicon cells*. These cells were the very first solar cells ever made for use in commercial applications. Inside the cell, the process of synthesis can result in the formation of a number of crystalline formations. It is less difficult to put into action in a production line. It is far less expensive than monocrystalline, despite the fact that it is less effective.

(3) *Amorphous solar cells*, which are neither organised nor crystallised at the molecular level (**Figure 5**), were traditionally used for smaller-scale applications because the term "amorphous," which means "shapeless" in English, gives the impression that they are.

(4) *Thin film Solar cells*—During the manufacturing process, several layers of photovoltaic material are layered on top of one another and stacked in the form of modules that are constructed of thin film solar cells. Using this process to develop a cell (thin film) results in a lesser requirement for silicon, which in turn results in cheaper production costs. In most cases, it is less expensive, but the rate of efficiency it achieves is significantly lower [7].

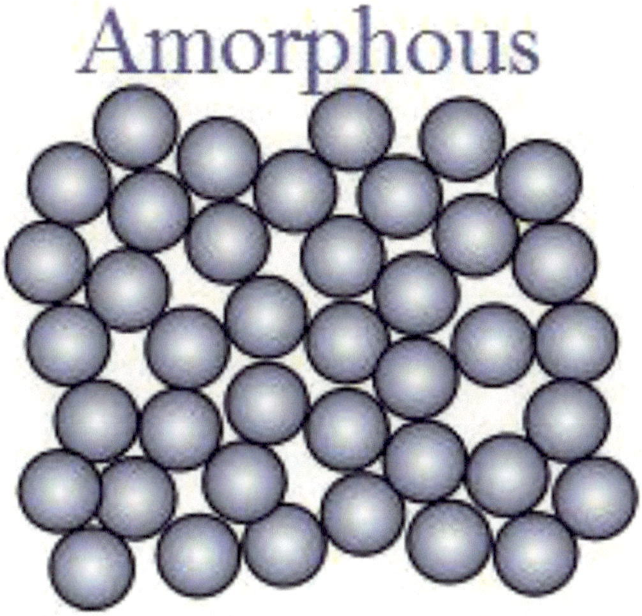

Figure 5.
Amorphous.

2.1.4 Equivalent circuit of a solar cell

In order to appreciate the electronic behaviour of a solar cell, it is helpful to construct a model that is electrically equivalent and is constructed on discrete electrical components whose behaviour is well understood. This model should be based on discrete electrical components. Because there is no such thing as a perfect solar cell in the actual world, the model incorporates both a shunt resistance component and a series resistance component (**Figure 6**). A current source that is coupled in series with a diode can successfully duplicate the characteristics of an ideal solar cell.

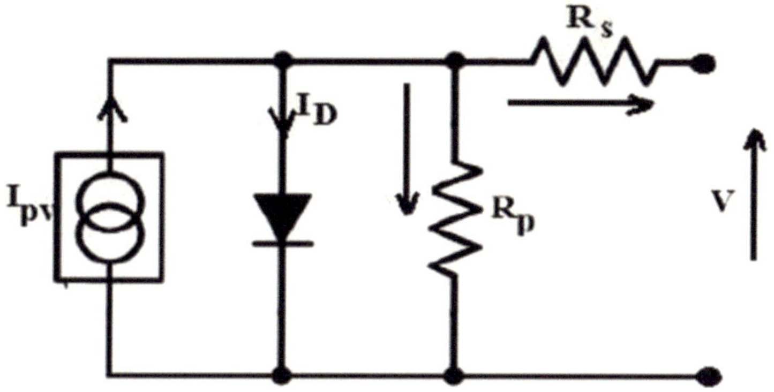

Figure 6.
The equivalent circuit of a solar cell.

2.1.5 Photovoltaic (PV) system

Solar cells are semiconducting devices that are responsible for converting solar energy into electricity. This transition takes place inside of a solar cell. A solar cell is a piece of equipment that can produce a specific amount of electrical power when exposed to sunlight. This particular kind of cell can produce both a voltage and a current as its output. A solar panel, which is also referred to as a photovoltaic module, is constructed out of a collection of solar cells and is used to generate electricity for the operation of various useful devices that require a particular voltage or current. Solar arrays are formed when individual solar panels are interconnected in order to generate huge amounts of solar-generated electricity.

The solar panels are just one component of a more extensive photovoltaic (PV) solar system. Depending on the application, the photovoltaic (PV) solar system may additionally comprise batteries for storing any excess electricity, dc/ac inverters for connecting the PV solar system to the electrical grid, as well as any additional electrical or mounting components. The balance of systems is the name given to the second component of the photovoltaic solar system, which is composed of these additional components (BOS). Last but not least, the solar system includes goods such as consumer electronics that can be powered by solar energy, such as radios and televisions. This shipment of products is the topic of our conversation.

The photovoltaic (PV) solar system is made up of a total of three components: (I) photovoltaic cells or solar arrays, (II) the system's overall equilibrium, and (III) the load.

2.2 Advantages of solar photovoltaic cells

1. Installing photovoltaic panels, which give clean, green electricity, is one of the most important advantages because it offers significant cost savings. As a result of the fact that there is no emission of potentially damaging greenhouse gases during the process of generating power using PV panels, solar PV is good for the environment.

2. Because solar energy is derived directly from the sun, it is an environmentally friendly, non-depletable, and economical source of energy. Solar power may be generated virtually wherever that there is access to sunlight. The power grid of the future will be a distributed power generation (DPG) system, and solar electricity is an excellent fuel source for this type of grid.

3. As a result of the rapid decline in the price of solar panels, which is forecast to continue over the next several years, the future of solar photovoltaic panels appears to be unusually promising in terms of both their economic viability and their environmental sustainability. Photovoltaic panels are a type of direct electricity generation that get their name from the fact that they generate power through photoelectric processes and are referred to as such.

4. When compared to the expenses of other types of renewable energy, it is often believed that the costs of operation and maintenance associated with photovoltaic panels are extremely cheap or nearly non-existent. Solar photovoltaic (PV) panels, with the exception of sun-tracking mechanical bases, do not contain

any mechanically moving elements. As a result, they break significantly less frequently and require much less maintenance than other energy sources that are in direct competition with them. Other energy sources include: (e.g., wind turbines).

5. Solar photovoltaic panels are perfect for usage in residential and urban areas because they do not produce any audible noise (see solar panels for homes).

6. Solar photovoltaic panels are an efficient approach for managing spikes in energy use, particularly during the scorching summer months, when the need for air conditioning is at its highest point. This is the time of year when energy consumption spikes the most.

7. Solar photovoltaic panels are one of the major renewable energy systems that are promoted through government subsidy funding (FITs, tax credits, etc.). As a consequence, the financial incentive for PV panels makes solar energy panels an attractive investment alternative, in spite of the fact that their prices have seen a significant reduction in recent years and are continuing to fall. In addition, the financial incentive for PV panels makes solar energy panels an attractive investment alternative.

8. Solar panels for residences may be installed on rooftops or in the ground in a rapid and easy manner, with no interference to regular activities.

2.3 Disadvantages of Solar photovoltaic cells

1. Solar energy, like other forms of renewable energy, has the potential to be unreliable due to factors such as the sun's inability to shine at night and the chance of cloudy or rainy skies during the day. Other forms of renewable energy do not have these limitations.

2. As a direct result, solar panels are not as reliable of a solution as other options due to the unpredictability and intermittent nature of solar energy.

3. An additional piece of hardware known as an inverter is required in order to connect solar panels to the power grid. This inverter is responsible for transforming the electricity from its direct current (DC) form into its alternating current (AC) form. This is important in order for the grid to be able to consume the electricity that has been generated.

4. In order to keep a consistent supply of electric power, solar panels require not only inverters but also storage batteries, particularly when they are connected to the grid. This is especially true when the grid is the source of the electric power. Investing in photovoltaic panels will therefore be more expensive as a result of this factor.

5. The installation of land-based PV panel systems necessitates the use of enormous tracts of land; in most cases, the land must be set aside for this purpose for a period of 15–20 years, and in some cases much longer.

6. When compared to the efficiency of other forms of renewable energy technology, solar panels' efficiency ranges from 14 to 25%, which is quite a bit lower than those other forms.

7. PV panels, despite the fact that they do not require much in the way of upkeep or operating costs, are nonetheless fragile and can be damaged with relative ease. Due to this fact, the additional insurance premiums that must be paid in order to safeguard a financial investment in photovoltaics (PV) are of the utmost significance.

2.4 Challenges and future perspectives

There are a number of problems that have a major impact on how well the PV system works. To start, there may be issues with data management during the PV system implementation. The proliferation of solar power plants has resulted in an explosion of data that is beyond the capacity of current technologies to manage. For this reason, the creation of a robust system capable of real-time monitoring of massive amounts of data is essential. The second crucial feature of a wireless monitoring system is its safety. The integrity of a system may be jeopardised due to inadequate security. Security measures must be followed to guarantee safety, privacy, and secrecy. Interference with the signals is a potential third issue. This can make it difficult to keep tabs on the solar PV system's statistics. As a result of signal interference, data transfer rates, module functionality, signal strength, and connectivity might all suffer. All of this results in lower service standards and greater revenue loss. In the fourth place, the node battery's lifespan determines the PV-overall system's energy efficiency. To be more specific, the PV network includes a number of different transmitter nodes. The longevity of the PV-network decreases with each failed node battery. To conclude, the programming language and operating system must be intuitive to use. Sixth, the data transfer range needs to be adequate for the specific PV system in question. Given that the distance between the furthest two PV panels in such a system can be measured in kilometres, this is of paramount importance. Data transmission speeds should not be an issue for a long-distance platform. Seventh, the dust, temperature, radiation, and humidity in the surrounding environment have a significant impact on the performance of the PV system. When installing PV panels in the open, special care must be taken to prevent damage. For instance, dust accumulation on a PV panel can lessen the amount of sunlight that is absorbed by the panel's solar cells. To add insult to injury (number eight), various electromagnetic and radio frequencies can impact the transmission module's accuracy. Thus, there is a need for shield-modules that are compatible with data transmission boards. The ninth problem is that the effectiveness of various solar cell methods varies substantially from one another and is extremely sensitive to the composition of the material from which the cells are made (e.g. mono-crystalline silicon, poly-crystalline silicon, amorphous silicon, thin-film copper indium disulfide, heterojunction incorporating thin film). As a result, picking the right solar cell technology is crucial for maximising output [8].

As solar photovoltaic (PV) systems grow in scale and area of use, future research should concentrate on improving data handling, boosting efficiency, increasing security, and expanding transmission range, and reducing signal interference. Specifically, you should avoid the nine issues we just went over. In reality, going forward, it will be necessary to create a trustworthy, cutting-edge method of monitoring wireless networks [8].

2.5 Application of Solar photovoltaic cell

Even though fossil fuels are still in high demand for automotive uses [9], photovoltaic systems have found their place in the market. A photovoltaic cell is a type of solar cell that generates electricity solely from the energy of incident light. A solar cell, also known as a photovoltaic cell, is a device that can convert light into electricity. Since PV cells may generate anywhere from a few kilowatts (KW) to enormous megawatts (MW), they have a much broader range of potential uses than traditional power plants. Only a select few will be explored briefly here:

1. It has been demonstrated that solar-powered water pumps are more effective in rural regions for the purpose of agricultural irrigation. Solar panels are used to generate electricity, which is then used to power a pump that moves water from a lower level to a higher level. One graphical representation of this idea is shown in the following figure for your perusal.

2. With regards to food preparation, solar cookers are currently all the rage. Solar cookers are readily available for purchase and use, with few maintenance requirements. The world's largest solar kitchen may be found at Taleti, which is located in the Indian state of Rajasthan at an elevation of 1219 m above sea level. It utilises solar steam cooking on six distinct modules and has 84 parabolic dish concentrators and shell-type receivers. Additionally, it has a total of 168 parabolic dishes [10].

3. Solar water and air heaters have been widely used for decades, long before PV cells were ever a possibility. The use of solar water heaters alone can significantly cut down on energy costs. These are effective in capturing the sun's rays, allowing us to use that heat to keep water at a comfortable temperature.

4. Solar panels are an easy way to power traffic lights in any location. The absence of shadows is the only real issue here.

5. Solar power can be utilised not only for conventional air conditioning systems but also for cold storage facilities. For these objectives, a vapour compressor system that makes use of solar photovoltaic panels and a vapour absorption system that makes use of thermal collectors are both viable options [11].

6. The Solar Photovoltaic (PV) System aboard the International Space Station Solar panels attached to the International Space Station provide an excess of electricity compared to what is needed to run the station [12]. The batteries of the space station are charged with around 60% of the power produced by the solar arrays when the station is exposed to sunlight. Solar panels on spacecraft offer power for two principal applications: [13] first, power to drive the sensors, active heating and cooling, and telemetry [14]. Solar panels on spacecraft also provide power for active heating and cooling [14]. Secondly, power for heating and cooling that is actively used. Then there is the power that is used for the propulsion of spacecraft, and that power is electric propulsion, which is also sometimes referred to as solar-electric propulsion. Solar panels made specifically for use in space are not the same as solar panels made specifically for use on Earth. They do not require any form of glass lamination, which means that they are resistant to moisture

even without it. The material that is used is distinct from other materials since it is required that the components be able to withstand high temperatures.

3. Conclusion

Damage to ecosystems and the natural equilibrium results from the use of fossil fuels. For the following reasons, solar energy may be the greatest long-term option: it is a huge and infinite energy source, its management has no negative effects on the environment, and it can be easily implemented in a variety of settings (from rural to urban to industrial). Thus, photovoltaic technology, which directly transforms sunshine into power, is incredibly practical. Especially considering how light and simple these gadgets are to operate. The photovoltaic system converts solar energy into usable electricity by activating electrons upon exposure to light. The PV-system has a long lifespan, functioning at 80% efficiency for 25 years and at 90% efficiency for 10 years. By allowing the PV panel's displacement to be adjusted in response to the sun's movement, efficiency can be increased. The PV system can be combined with both renewable (such as wind) and non-renewable (such as diesel) energy sources to provide reliable power. They are examples of what are known as hybrid systems. Although there are many positive aspects of solar energy production, there are also some disadvantages to consider. These include the upfront cost, the need for regular maintenance (which can have an environmental impact), and the challenge of raising consumer awareness. Nonetheless, we must be conscious of the fact that solving these issues is the path to environmental conservation around the world.

Acknowledgements

The author is greatly indebted to prof. Dr. Vishal Ramchandran Panse for his kind support and encouragements. His valuable suggestions are highly appreciated.

Conflict of interest

The authors declare no conflict of interest.

Solar Photovoltaic Principles
DOI: http://dx.doi.org/10.5772/intechopen.109730

Author details

Aparna Dixit[1*], Arti Saxena[2], Ramesh Sharma[3], Debidatta Behera[4] and Sanat Mukherjee[4]

1 Department of Basic Science and Humanities, Pranveer Singh Institute of Technology, Kanpur, Uttar Pradesh, India

2 Department of Electronics and Communication, Pranveer Singh Institute of Technology, Kanpur, Uttar Pradesh, India

3 Department of Basic Science and Humanities, Feroz Gandhi Institute of Engineering and Technology, Raibareily, Uttar Pradesh, India

4 Department of Physics, Birla Institute of Technology, Mesra, Ranchi, Jharkhand, India

*Address all correspondence to: draparnadixit1@gmail.com

References

[1] International Renewable Energy Agency. Renewable Capacity Statistics 2022 [Statistiques De Capacité Renouvelable 2022 ESTADÍSTICAS De Capacidad Renovable 2022]. 2022

[2] Communications Materials—United Nations Sustainable Development. 2022. Available from: https://www.un.org/sustainabledevelopment/news/communications-material/

[3] Available from: https://www.smithsonianmag.com/innovation/inside-first-solarpowered-flight-around-world-180968000/

[4] Solar PV Power Generation in the Net Zero Scenario, 2000-2030—Charts—Data and Statistics—IEA. Available from: https://www.iea.org/data-and-statistics/charts/solar-pv-power-generation-inthe-net-zero-scenario-2000-2030 [Accessed: May 3, 2022]

[5] Wolf S et al. Towards industrial application of isotropic texturing for multi-crystalline silicon solar cells. In: Proc. of the 16th European PV Solar Energy Conference. Glasgow, United Kingdom. 2000

[6] Davidson J. The New Solar Electric Home. 2001

[7] The Renewable Energy Hub. Available from: https://www.renewableenergyhub.co.uk/ [Accessed: May 25, 2022]

[8] Ansari S, Ayob A, Hossain Lipu SM, Hanif MD, Saad M, Hussain A. A review of monitoring technologies for solar PV systems using data processing modules and transmission protocols: Progress, challenges and prospects. Sustainability. 2021;**13**:8120. Available from: https://bit.ly/3rX1mOD

[9] North Carolina Solar Center. Photovoltaic Applications

[10] World's Largest 38500-meal Solar Kitchen in India. Inhabitat. 2014

[11] Naganagouda H. A Complete Handbook of Solar Energy. Bangalore: K.P.C.L.

[12] Available from: https://www.nasa.gov/mission_pages/station/structure/elements/solar_arrays.html#.WXY9i4h95PY

[13] NASA JPL Publication. Basics of Space Flight, Chapter 11. Typical Onboard Systems, Propulsion Subsystems. Available from: http://www2.jpl.nasa.gov/basics/bsf11-4.html#propulsion

[14] Available from: http://www.alternative-energy-tutorials.com/solar-power/stand-alone-pv-system.html

Chapter 2

Transparent Solar PV Panels

Tulshi Shiyani

Abstract

Global warming is increasing emissions of greenhouse gases. It damages the environment of Earth. Solar energy is the cleanest source of renewable energy. It is an abundant source of clean energy. It has tremendous scope to generate electricity. Solar cells are devices that convert solar energy into electrical energy. Transparent solar panels are made up of transparent solar cells or *transparent luminescent solar concentrators*. A transparency of about 80% has been achieved with power conversion efficiency of about 12–15% in transparent solar cells. These cells can be used in buildings, vehicles, and other desired applications to generate solar power. We discuss solar energy basics and its conversion technologies. Transparent solar panels may bring a revolution in low-power display devices and mobile applications.

Keywords: solar energy, solar cells, transparent solar panel, electrical energy, renewable energy

1. Introduction

1.1 Conventional solar panels

Solar energy is the cleanest source of renewable energy on earth. Solar cell works on the principle of photoelectric effect that converts solar energy into electrical energy. Solar cells are mainly categorized into four generations. The conventional mono and polycrystalline wafer-based solar cells are first-generation [1, 2]. Thin films solar cells such as CIGS, CdTe, and CZTS are second-generation of solar cells. Multijunction solar cells are third-generation solar cells. Quantum dot and hybrid solar cells are fourth generation of solar cells. Theoretically, each generation of solar cells has energy conversion efficiency. Experimentally, multijunction solar cells have achieved the highest energy conversion efficiency, about 47.5%. Silicon solar cells are mostly commercialized technology due to abundant and cheap materials. However, its fabrication involves expensive and toxic processes. Currently, solar panels are used as off-grid and rooftop on-grid applications. Therefore, novel solar panel technology is required for multiple applications such as windows, displays, surface, etc. [3–5].

1.2 Transparent solar panels

A transparent solar panel is a basically challenging idea because sunlight (photons) must be absorbed by solar cells and converted into electrical energy (electrons). Sunlight can pass through the medium in transparent solar glass and it defeats the conversion purpose. But in transparent solar panels, the absorption happens in a different way. The cell selectively harnesses a portion of the sunlight that is invisible to naked eye and allows the visible light to pass through the device. The researchers have developed the transparent luminescent solar concentrator (TLSC) to achieve the transparent behavior of the cell rather than trying to develop the challenging transparent PV glass cell [6, 7]. Transparent solar panels use transparent luminescent solar concentrators as glass, which is transparent in nature. It uses organic molecules to absorb invisible spectrum of sunlight. So, these organic molecules absorb the specific IR and UV light. The electrons of molecules are excited by the energy of radiation and jump to a higher-level orbit. When they settle at ground state, the energy is released in the form of IR radiation (luminescent or glowing) of the different non-visible spectrum. This emitted IR light is guided through the edge of the plastic, where the strips of traditional solar cells convert it into electrical energy. This technology is feasible on cars windows, buildings, mobile phones, and other devices with a transparent surface. The limiting factor of this technology for commercialization is its low efficiency of about 1%. However, efficiency can be achieved by tuning the properties of the material in device. The first fully transparent solar concentrator was built by researchers at Michigan State University in 2014. This transparent solar panel could turn any glass sheet or window into a photovoltaic cell. The full transparency was achieved for the solar glass by 2020. Transparent solar panel technologies are set to transform the solar energy utilization landscape globally. We may able to generate electricity from windows of building, vehicles, phones, etc. These transparent solar panels can be deployed easily in various devices and systems such as laptops, e-readers, skyscrapers, windows, etc. The glass windows in buildings can be replaced by solar power windows [8, 9].

The organic salts absorb IR and UV light and emit IR in case processes occur outside the visible spectrum. Thus it appears transparent. TLSC is composed of organic salts that absorb specific UV and IR light, which then luminesce (glow) as another invisible wavelength. This wavelength is then guided to the edge of the window-thin PV solar cell strips that convert it into electricity. The mass production of such transparent solar panels can give an efficiency of about 10%. This can help to generate power through every window of home or office buildings and may bring a transformative result [10].

Transparent solar panels are categorized into (i) partially transparent solar panels: Heliatek Gmb, a German manufacturer has developed this technology that can absorb 60% of the sunlight it receives. They have achieved an energy conversion efficiency of about 7.2%. However, the generation of solar power can be increased by adjusting the transmitted and absorbed sunlight, for example, south-facing glass buildings can reduce the transmitted light. (ii) fully transparent solar panels: The researchers at MSU, USA have achieved fully transparent PV glass panels that resemble regular glass. They have achieved energy conversion efficiency of about 10%. Bigger or more windows can generate more solar power [11–13].

Currently, researchers at Michigan State University and MIT as well as manufacturers such as Brite solar, Physee, and Ubiquitous energy are pioneers in transparent solar panel technology as shown in **Figure 1**. Ubiquitous energy has achieved energy conversion efficiency of 9.8% and working toward developing net-zero energy buildings [14, 15]. Physee has introduced power windows that work as building blocks for smartskin. Smartskin can work for sensing, power generation, and regulating the inside climate using an artificial intelligent system. Richard Lunt has proposed making a solar cell that would absorb all the energy from the sun except the part that allows us to see. He has developed highly transparent solar cells that represent the

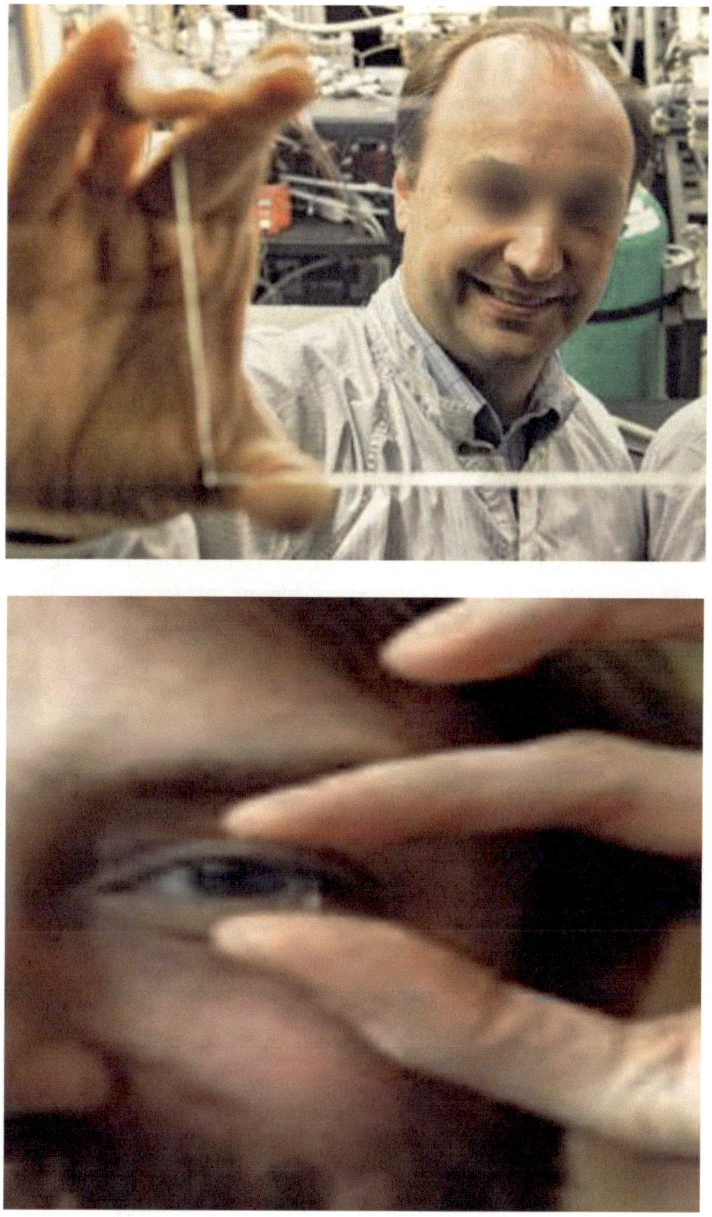

Figure 1.
Vladimir Bulović of electrical engineering and computer science showing their transparent solar cells (upper), and Richard Lunt demonstrates the transparency of the novel solar cell at MIT (lower).

wave of the future. He has claimed that these solar panels have similar power generation potential as rooftop. This can also be applicable to buildings, cars, mobile, and other such devices. MIT researchers are developing transparent solar cells that could turn every product such as windows and electronic devices into power generators. These cells can absorb only IR and UV light. They have developed a room-temperature fabrication method and can deposit coatings of solar cells on various materials to run electronic displays using ambient light [16–18].

Solar Panel blinds are accessories to transparent solar glass or panels in case of windows for generating electricity. This blocks direct sunlight from entering inside. SolarGaps introduce solar blinds and claim that they can generate about 100 W of power on every 10 sq. ft. of window area. It can be installed from outside or inside and can control its angle according to the sun's position [19, 20].

2. Device configuration of transparent photovoltaic device

The schematic representation of a transparent photovoltaic device is shown in **Figure 2**. This shows the key component of transparent solar cell, which transmits visible light and captures NIR and UV light. The thick layer is glass, plastic, or other transparent material. The coating of PV materials on top of the device. These PV materials are photoactive layers. The active layers include semiconducting materials that get excited upon falling sunlight and interact, creating an electric field that causes electric current to flow in a device through a circuit. This current can be taken out of the device to an external circuit via connecting sandwiched electrodes. Both electrodes must be transparent and they are anti-reflective coatings to reduce light reflection. Therefore, a combination of optical design, molecular engineering, and optimization of the device are used to design transparent PV devices [21].

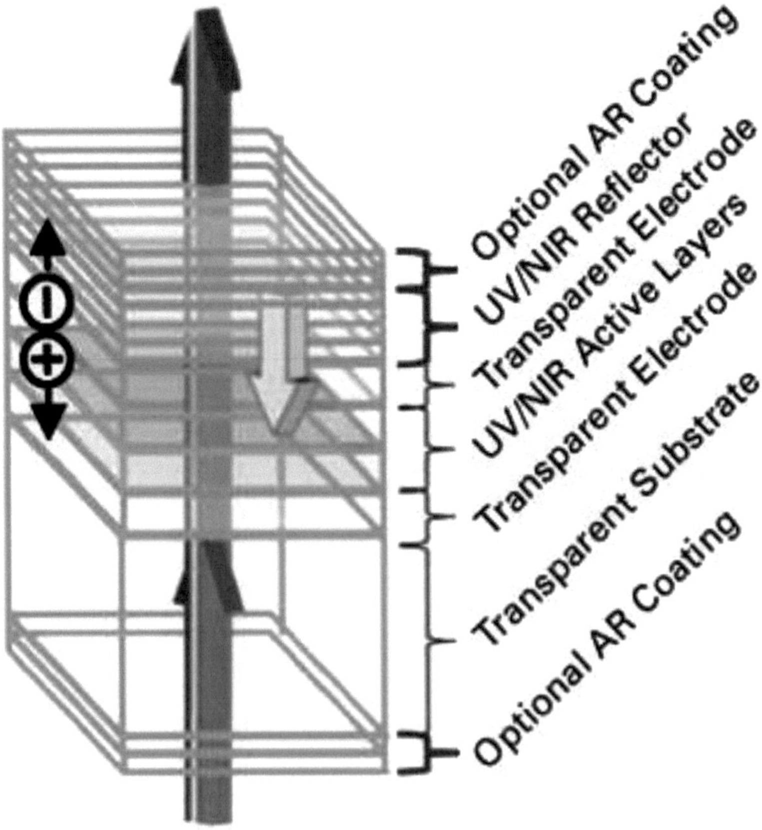

Figure 2.
Schematic of transparent photovoltaic device.

3. Spectral response of conventional and transparent PV cells

The spectral response of conventional and transparent solar cells is shown in **Figure 3**. The absorptive response (black curve) is superimposed on the solar spectrum (gray curve). In the conventional cell, the wavelengths at which absorption is relatively high include the visible spectrum (400–700 nm). The transparent cell absorbs well in the near-infrared and the ultraviolet spectrum but the absorption drops off and approaches zero in visible range [22].

The current transparent solar cells transmit more than 70% of the visible light, which is within the range of tinted glass used in the windows of buildings with a

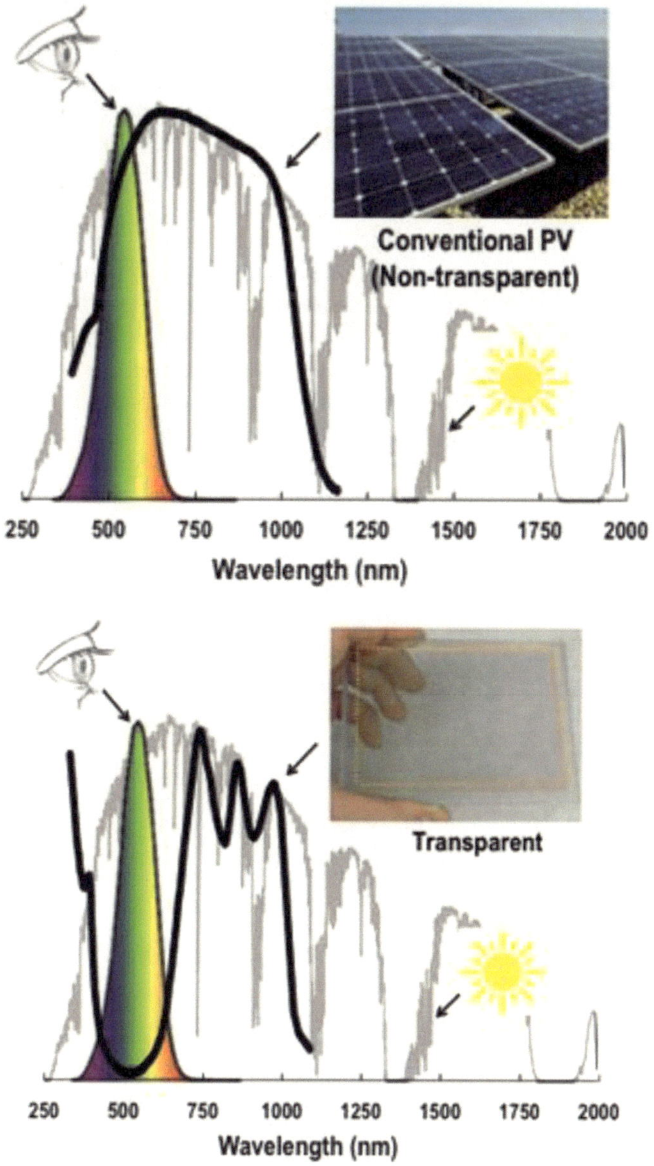

Figure 3.
Spectral response for conventional (upper) and transparent PV devices (bottom).

power-conversion efficiency of about 2%. Lunt and Bulović claimed that it can able to reach over 12% efficiency on basis of theory. Lunt already has demonstrated transparent cells integrated into series can power the liquid crystal display on a small clock, relying entirely on ambient light [23].

4. Transparent flexible solar cells

A flexible transparent solar cell was first developed at MIT as shown in **Figure 4**. Researchers have developed a novel technique using graphene to prepare solar cells on surfaces of glass, plastic, paper, and tape. This device contains carbon-based organic materials with graphene electrodes and transparent material. This device involves a layer of graphene on the solar cell. The flexible transparent solar cells would be lightweight and cheap technology as well as low-cost materials [23]. Organic materials absorb the UV and IR components of the solar spectrum but transmit visible light. The most affordable option is to use indium tin oxide (ITO) coated flexible substrate. But ITO is brittle and may break during deposition on flexible substrate. A one-atom thick layer of graphene can be developed as alternative to ITO in transparent solar cells. This alternative material is flexible, highly conductive, and transparent. The most important thing is that it is made up of inexpensive carbon material [24].

There are some problems during the deposition of graphene electrodes on solar cells. The first problem is depositing 1 atom layer thick of graphene. A bottom layer of graphene is deposited directly on the substrate and then the top layer. The fabrication of graphene on the top layer in solar cells as a hole transport layer is very tricky. The other main problem is preparing graphene electrodes for different applications. One electrode should let electrons flow out easily in solar cells. So, here both graphene electrodes have different work functions. Therefore, changing the work function is not an easy task. A layer of ethylene-vinyl acetate (EVA) can be incorporated into the graphene layer to make the device more flexible. This would differ both graphene electrodes to work for different purpose [25].

In practical, a transparent solar cell was developed using graphene, ITO, and aluminum materials. The performance of this solar cell was lower than solar cells with one aluminum electrode. Aluminum electrode on the bottom reflects some incoming light back into solar cell and increases the absorbance of sunlight more than a

Figure 4.
Transparent solar cell on flexible substrate developed at MIT.

transparent cell. The PCE for graphene/graphene devices was comparable to existing commercial solar panels, about 4.1% with transparency of about 61%.

The organic solar cell has the advantage to deposit on any type of surface, such as rigid or flexible, transparent, etc. The graphene/graphene devices have been demonstrated on various flexible substrates such as plastic, opaque paper, translucent kapton tape, etc. The performance was nearby in all these devices [26].

5. Manufacturing of transparent solar panels

Polysolar is developing transparent solar panels for buildings, canopies, and greenhouses applications. These panels can be used on walls with non-directional ambient sunlight. They have installed transparent solar panels in some buildings and bus shelters. This has allowed manufacturing of PV powering interactive displays, lighting, and signage. They have achieved power conversion efficiency of up to 12–15% in gray-tinted panels. Polysolar is trying to cover maximum area for generation of solar power and increase the transparent PV footprint in various sectors [27].

Thin-film PV cells (in *orange color*) are deposited as a naturally translucent layer onto the glass (*blue*) before another glass sheet is laminated on top as shown in **Figure 5**. They allow the panels to work at a much higher efficiency at varying angles than regular solar panels and can generate solar power at low levels of sunlight [28–30].

Ubiquitous energy is also developing transparent PVs using semiconductor with higher efficiencies and transparencies. They are also trying to assemble panels on electronic devices such as mobile phone display to self charge the devices.

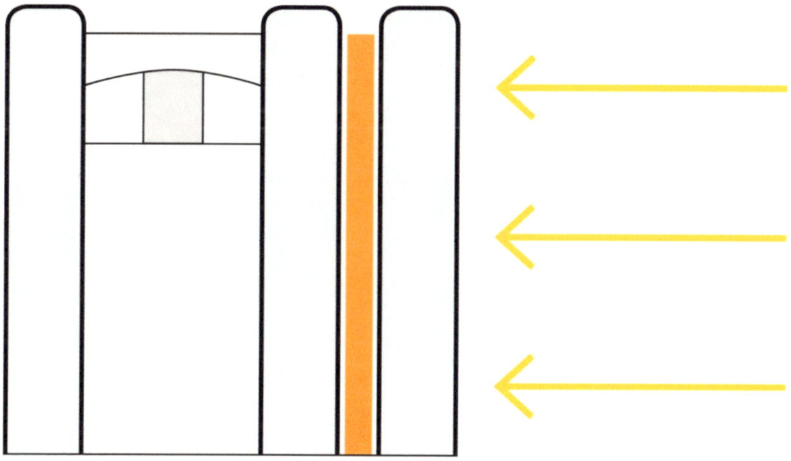

Figure 5.
Thin-film photovoltaic cells (in orange) are deposited as a naturally translucent layer on to the glass (blue).

6. Conclusion

Transparent solar panels are very challenging to find their real time applications. They have huge scope to generate solar power and making electronic devices for self charging applications. Currently, they have limitations on large applications due to low efficiency and cost. These solar cells can block the most of IR. Researchers have developed graphene based organic solar cells with improved efficiency with significant transparency. Increasing the amount of active area would push up the PCE, but transparency would drop. The highest transparency of about 80% has been measured in transparent solar cells with maximum power conversion efficiency of about 12–15%. The various materials have been utilized to improve the transparency and performance of solar cells. Transparent solar panel would bring a remarkable change in electronic and optical applications.

Acknowledgements

The authors are thankful to Shiyani Research Institute, Rajkot, and Shiyani Enterprise (OPC) Private Limited, Rajkot for providing technical support to publish this chapter.

Author details

Tulshi Shiyani
Saurashtra Research Institute, SRI, Rajkot, India

*Address all correspondence to: sh.ts@protonmail.com

References

[1] Lunt RR, Bulović V. Transparent, near-infrared organic photovoltaic solar cells for window and energy-scavenging applications. Applied Physics Letters. 2011;**98**:113305. DOI: 10.1063/1.3567516

[2] Lunt RR, Osedach TP, Brown PR, Rowehl JA, Bulović V. Practical roadmap and limits to nanostructured photovoltaics. Advanced Materials. 2011;**23**:5712-5727. DOI: 10.1002/adma.201103404

[3] Available from: https://www.solarmango.com/2016/09/25/can-transparent-solar-panels-big-boost-solar-tall-buildings/

[4] Da Y, Xuan Y. Li Q, From light trapping to solar energy utilization: A novel photovoltaic-thermoelectric hybrid system to fully utilize solar spectrum. Energy. 2016;**95**:200-210. DOI: 10.1016/j.energy.2015.12.024

[5] Atul T, Maheshwar S. Solar Cell Nanotechnology. New Jersey: Wiley; 2013

[6] Hosenuzzaman M, Rahim NA, Selvaraj J, Hasanuzzaman M. Global prospects, progress, policies, and environmental impact of solar photovoltaic power generation. Renewable and Sustainable Energy Reviews. 2015;**41**:284-297. DOI: 10.1016/j.rser.2014.08.046

[7] Fahrenbruch A. Photovoltaic Solar Energy Conversion. California: Elsevier; 2012

[8] Tyagi VV, Rahim NAA, Rahim NA, Selvaraj JAL. Progress in solar PV technology: Research and achievement. Renewable and Sustainable Energy Reviews. 2013;**20**:443-461. DOI: 10.1016/j.rser.2012.09.028

[9] Parida B, Iniyan S, Goic R. A review of solar photovoltaic technologies. Renewable and Sustainable Energy Reviews. 2011;**15**:1625-1636. DOI: 10.1016/j.rser.2010.11

[10] Green MA, Emery K, Hishikawa Y, Warta W. Solar cell effic tables. Photovoltaics. 2011;**2011**:84-92. DOI: 10.1002/pip.1088

[11] Rehman S, Bader MA, Al-Moallem SA. Cost of solar energy generated using PV panels. Renewable and Sustainable Energy Reviews. 2007;**11**:1843-1857. DOI: 10.1016/j.rser.2006.03.005

[12] Bruton TM. General trends about photovoltaics based on crystalline silicon. Solar Energy Materials & Solar Cells. 2002;**72**:3-10. DOI: 10.1016/S0927-0248(01)

[13] Braga AFB, Moreira SP, Zampieri PR, Bacchin JMG, Mei PR. New processes for the production of solar-grade polycrystalline silicon: A review. Solar Energy Materials & Solar Cells. 2008;**92**:418-424. DOI: 10.1016/j.solmat.2007.10.003

[14] Muller A, Ghosh M, Sonnenschein R, Woditsch P. Silicon for photovoltaic applications. Materials Science Engineering. 2006;**134**:257-262. DOI: 10.1016/j.mseb.2006.06.054

[15] Mirzaei M, Zamani M. Energy for sustainable development a comparative analysis of long-term fi eld test of monocrystalline and polycrystalline PV power generation in semi-arid climate conditions. Energy for Sustainable Development. 2017;**38**:93-101. DOI: 10.1016/j.esd.2017.01.002

[16] Zweibel K. Thin film PV manufacturing: Materials costs and their

optimization. Solar Energy Mater Sol Cells. 2000;**63**:375-386. DOI: 10.1016/S0927-0248(00)00057-X

[17] Hamza A, Ali H, Abdelrasheed H, Zeid S, Alfadhli HMG. Energy performance, environmental impact, and cost assessments of a photovoltaic plant under Kuwait, climate condition. Sustain Energy Technol Assess. 2017;**22**:25-33. DOI: 10.1016/j.seta.2017.05.008

[18] Günes S, Neugebauer H, Sariciftci NS. Conjugated polymer-based organic solar cells. Chemical Reviews. 2007;**107**:1324-1338. DOI: 10.1021/cr050149z

[19] Conibeer G, Green M, Corkish R, Cho Y, Cho EC, Jiang CW, et al. Silicon nanostructures for third generation photovoltaic solar cells. Thin Solid Films. 2006;**511-512**:654-662. DOI: 10.1016/j.tsf.2005.12.119

[20] Conibeer G. Third-generation photovoltaics. Materials Today. 2007;**10**:42-50

[21] Fraas L, Partain L. Solar Cells and their Applications. Second ed. John Wiley & Sons, Inc.; 2010. ISBN: 9780470636886. DOI: 10.1002/9780470636886

[22] McEvoy AJ, Castaner L, Markvart T. Solar Cells: Materials, Manufacture and Operation. 2nd ed. Oxford: Academic Press; 2012. ISBN: 9780123869647

[23] Zhang HL, Van Gerven T, Baeyens J, Degrève J. Photovoltaics: Reviewing the European feed-in-tariffs and changing PV efficiencies and costs. Scientific World Journal. 2014;**2014**:1-10. Article ID 404913. DOI: 10.1155/2014/404913

[24] Green MA. Solar Cells: Operating Principles, Technology, and System Applications. Englewood Cliffs, NJ: Prentice-Hall, Inc; 1982

[25] Hamadanian M, Safaei-Ghomi J, Hosseinpour M, Masoomi R, Jabbari V. Uses of new natural dye photosensitizers in fabrication of high potential dye-sensitized solar cells (DSSCs). Materials Science in Semiconductor Processing. 2014;**27**:733-739. DOI: 10.1016/j.mssp.2014.08.017

[26] O'Regan B, Grätzel M. A low-cost, high-efficiency solar cell based on dye-sensitized colloidal TiO2 films. Nature. 1991;**353**:737-740. DOI: 10.1038/353737a0

[27] Hamadanian M, Jabbari V, Gravand A. Dependence of energy conversion efficiency of dye-sensitized solar cells on the annealing temperature of TiO2 nanoparticles. Material Science Semiconductor Process. 2012;**15**:371-379

[28] Huang ZS, Hua T, Tian J, Wang L, Meier H, Cao D. Dithienopyrrolobenzotriazolebased, organic dyes with high molar extinction coefficient for efficient dye-sensitized solar cells. Dyes and Pigments. 2016;**125**:229-240. DOI: 10.1016/j.dyepig.2015.10.022

[29] Gong J, Sumathy K, Qiao Q, Zhou Z. Review on dye-sensitized solar cells (DSSCs):Advanced techniques and research trends. Renewable and Sustainable Energy Reviews. 2017;**68**:234-246. DOI: 10.1016/j.rser.2016.09.097

[30] Chen JG, Chen CY, Wu SJ, Li JY, Wu CG, Ho KC. On the photophysical and A.A.F. Husain et al. Renewable and Sustainable Energy Reviews. 2018;**94**:779-791

Chapter 3

Photovoltaic and Photothermal Solar Cell Design Principles: Efficiency/Bandwidth Enhancement and Material Selection

Shiva Hayati Raad and Zahra Atlasbaf

Abstract

There are two main approaches for developing solar cells, including photovoltaic and photothermal technologies. Photovoltaic solar cells benefit from an active region whose performance can be improved by embedding nanoparticles with different shapes and materials. Photothermal solar cells are broadband absorbers, enabling electromagnetic energy absorption in the solar radiation region. Since the solar spectrum is expanded from 120 to 1000 THz, the device bandwidth engineering and its efficiency enhancement through utilizing nanoparticles, multiresonance configurations, and multilayered structures are necessary. Moreover, using chemically inert materials with high thermal conductivities results in stable performance under different environmental conditions. Thus, in this chapter, various photovoltaic and photothermal solar cells will be discussed, emphasizing their design principles. The chapter mainly considers absorption bandwidth enlargement, absorption efficiency enhancement, and material selection considerations. In this regard, solar cells designed with plasmonic materials, transition metals, refractory metals, and carbon materials are presented. Notably, the potential of two-dimensional graphene material in the solar cell design is revealed, and a lightweight graphene-based solar cell with near-perfect coverage of the whole solar spectrum is introduced.

Keywords: photothermal solar cell, photovoltaic solar cell, bandwidth enhancement, efficiency, material selection, 2D materials

1. Introduction

Solar energy has been known as the most important renewable energy source in the world. Thus, designing devices for solar radiation storage as electricity (photovoltaic solar cell) or thermal energy (photothermal solar cell) is of great interest. Although the latter technology requires a further energy conversion process to produce electricity from harvested thermal energy, the conversion can be achieved with high efficiency. To design solar cells, different types of materials are used in geometrically engineered configurations, each having its pros and cons. The important parameters for evaluating solar cells are their efficiencies, bandwidth, tolerance to environmental conditions, and robustness to the incident angles of incoming waves [1, 2].

The photovoltaic solar cell design can be achieved by employing thin film technology (efficiency of 23.4%), multijunction devices (39.2% efficiency), crystalline silicon (c-Si) based configurations (theoretical efficiency of 26.7%), perovskite (theoretical efficiency of 31%), organic thin films (16.4% efficiency), dye-sensitized method (12.3% efficiency), and perovskite-based quantum dot usage (16.5% efficiency) [3–6]. Light trapping and confinement capabilities provided by the plasmonic nanoparticles play a crucial role in improving the efficiency of photovoltaic solar cells [7]. Moreover, by using wrinkle-like graphene sheets over the plasmonic nanoparticle, the photocurrent density enhancement can exceed the light trapping limit of the textured screens due to the broadband absorption of bend carbon [8]. Monolayers of semiconducting transition-metal dichalcogenides (TMDs) such as MoS_2 are also promising candidates for absorption efficiency enhancement in Si-based photovoltaics [9].

Photothermal solar cells are electromagnetic wave absorbers, and there is numerous research in the literature dealing with the electromagnetic absorber design in any desired frequency. The design principles of the absorbers obey the same roles, regardless of the selected spectrum. Thus, reviewing the wideband absorber design methods may be beneficial for establishing ideas for efficient solar cell design. The absorption rate of the thermal absorbers is directly connected to their ability to block wave transmission and eliminate wave reflection. This type of solar cell is commonly designed with an array of elements, and the performance of the anti-reflection coating depends on the shape of the constructing elements [10].

Microwave pyramidal absorbers are widely used broadband absorbers, in which the tapered nature of the geometry results in bandwidth enhancement by fulfilling the above-mentioned conditions [11]. The same geometry has been used as an efficient wideband absorber in the millimeter wave spectrum [12]. The pyramidal texture has also applications in solar cell design with high efficiency and industry standards [13]. Moreover, designing a device with fourfold symmetry has a great influence on the polarization in-sensitivity of the solar cell, and solar cells with high tolerance against the incident angle of the incoming wave are desired. Importantly, designing a device that operates under different environmental conditions, including temperature and humidity, increases the reliability of the system. Material selection plays an important role in this regard. In this chapter, photovoltaic and photothermal solar cell technologies will be introduced. Later, different methods of improving their performance (efficiency and bandwidth) are discussed. Due to the importance of material selection in solar cell performance, guidelines for choosing the proper material combinations are presented in detail. Finally, the remarkable properties of two-dimensional graphene material in the full-spectrum solar cell design are revealed.

2. Solar cell design technologies

Two well-known solar cell design technologies employ the photovoltaic or photothermal mechanism for light harvesting. The first method directly converts the absorbed solar energy into electricity. In the second method, the absorbed energy is of the thermal type, and it is later converted into electrical energy. **Figure 1a** shows a photovoltaic solar cell, in which n and p-type doped semiconductors are used in the

DOI: http://dx.doi.org/10.5772/intechopen.110093

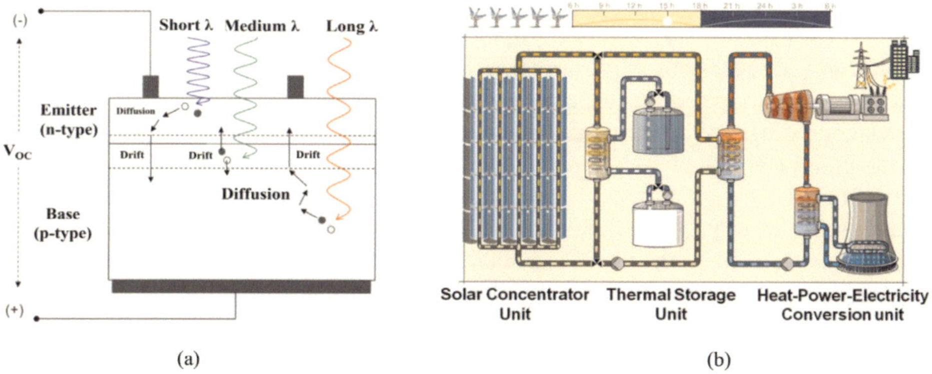

Figure 1.
(a) Photovoltaic solar cell and (b) electric energy production using photothermal solar cell [14].

design. When the sun's photons hit this structure, they are absorbed and create energy carriers (electrons and holes) that contribute to producing electrical energy [7]. Moreover, **Figure 1b** shows the performance of a photothermal solar cell in which mirrors and lenses are used to focus solar energy. This thermal energy can run old steam turbines or Stirling engines to generate electricity. The critical point is that the heat energy, collected by the advanced thermal solar cells, can be stored and used to produce electrical power at the proper time [14].

It is interesting to mention that photovoltaic solar cells cannot absorb photons at wavelengths higher than their bandgap. Thus, a combination of two technologies can be a solution for designing effective solar cells [14]. On the other hand, a thermophotovoltaic system enhances absorption efficiency during the dark hours, and it consists of a heat source (sun), thermal absorber, and photovoltaic solar cell. The theoretical conversion efficiency of this pollution-free and portable system with a low-maintenance cost is 85.4% [15]. For instance, by using germanium selenide (GeSe) nano-pyramids with optimized heights in the active layer of the perovskite solar cell, the light at a wavelength higher than 800 nm can also be absorbed [16]. As another approach, the exploitation of the different III-V semiconductors separately generates electron-hole pairs in different parts of the incoming solar spectrum for wider absorption coverage [17].

There are three generations of photovoltaic solar cells in the market. The first generation is based on crystalline silicon and has a high conversion efficiency. The high cost of this generation, due to the required large material thickness, is its main drawback. The second generation aims to reduce the device cost without scarifying efficiency. In this category, thin-film cells have gained lots of interest due to the reduction in the utilized raw materials [7, 18, 19]. They usually include a metallic contact layer, an active region, and an indium tin oxide (ITO) top layer. Thin film solar cells collect the carriers effectively, but they suffer from a low absorption rate due to small optical path length. The third generation of solar cells aims to enhance the efficiency of the second generation, and it is an ongoing research topic [20, 21]. An essential parameter of photovoltaic solar cells is their quantum efficiency, which is calculated as the ratio of absorbed power to incident power. Thus [22],

$$QE(\lambda) = \frac{P_{abs}(\lambda)}{P_{in}(\lambda)} \tag{1}$$

The high quantum efficiency of the absorber shows that when the solar cell is exposed to a photon with an arbitrary wavelength, it can generate a significant current.

Photothermal solar cells exploit different patterns on top of the engineered substrates to absorb solar energy. **Figure 2a** shows an instance of this absorber category. The efficiency of the photo thermal solar cell is strongly dependent on the efficiency of the designed electromagnetic wave absorber [23, 24]. Apart from the high absorption rate, the operating bandwidth coverage (120–1000 THz) is a critical point to tarp the whole solar energy. To evaluate the performance of the solar cell, its absorption spectrum is compared with the AM1.5 solar radiation spectrum, as in **Figure 2b** [23]. The short circuit current density of the solar energy for AM1.5 is defined as [25]:

$$J_{SC} = \frac{e}{hc} \int_{\lambda_{\min}}^{\lambda_{\max}} \lambda \Phi_{AM1.5}(\lambda) A(\lambda) \tag{2}$$

where e is the electron charge, h is the Planck's constant, c is the speed of light, λ is the wavelength, $\Phi_{AM1.5}$ is solar radiance at AM 1.5, and A is the absorption.

3. Absorption efficiency/bandwidth enhancement in solar cells

This section aims at introducing different approaches for absorption efficiency/bandwidth enhancement in both photovoltaic and photothermal solar cells.

3.1 Absorption efficiency enhancement

Nanoparticle arrays with different shapes, materials, and the number of layers are usually used for absorption efficiency enhancement [22, 26–28]. To evaluate the absorption rate in the photothermal solar cells, the unit cell analysis with Floquet port excitation is commonly exploited. The absorption rate (A) is related to the reflection coefficient (R) and transmission coefficient (T) as A = 1-R-T. **Figure 3a** shows a graphene-based plasmonic nanoparticle array residing on top of the metallic reflector, and **Figure 3b**

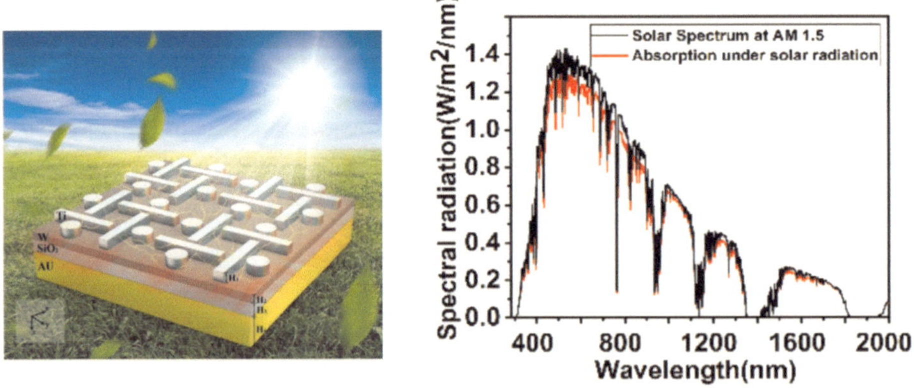

(a) (b)

Figure 2.
(a) Illustration of the photothermal solar cell and (b) comparison of its spectral radiation with AM1.5 solar spectrum [23].

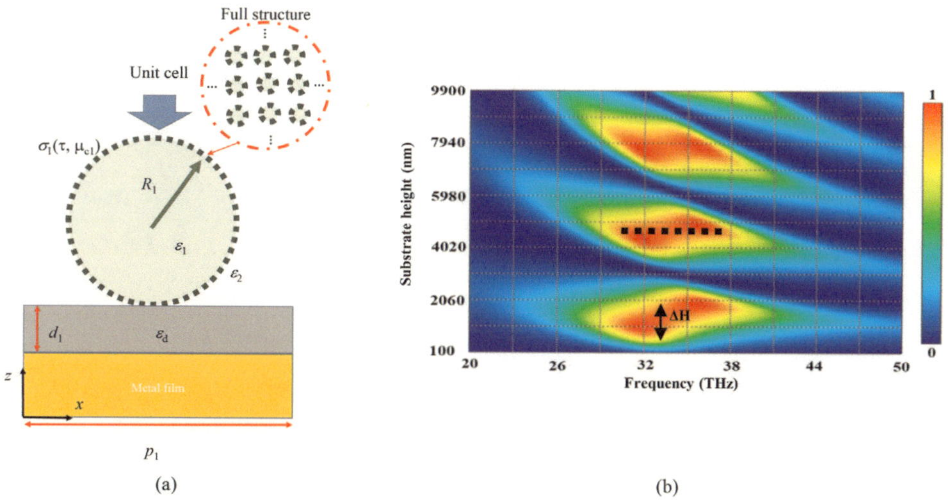

Figure 3.
(a) Absorption efficiency engineering using a back reflector and (b) absorption rate of the device in terms of substrate height [29].

shows the absorption rate of the device in terms of substrate height. The perfect absorption rate of the device is due to the excitation of localized surface plasmon resonances in the spherical graphene shells [29].

Although the operating frequency of the provided example is not in the solar spectrum, the method is general and has been widely used for absorption enhancement in different types of solar cells. For instance, silver nanorods are used as the back electrode in a thin film solar cell to achieve a 45% improvement in conversion efficiency [27]. Also, gold and silver spherical nanoparticles are embedded in the rear layer of a perovskite solar cell to improve its efficiency in the red region of the visible spectrum [30]. By investigating the role of dispersion and dissipation of the nanoparticle on the performance, it is proved that dielectric nanoparticles lead to higher enhancement compared to their metallic counterparts [31]. Moreover, the metals are oxidized in various weather conditions, and the structure's performance is greatly affected by the oxide layer [32]. Thus, high-index dielectric nanoparticles (e.g., titanium dioxide), supporting magnetic Mie resonances, are used to transform the freely propagating sunlight into guided modes (**Figure 4a**) [33]. Also, silica sphere, hemisphere, moth-eye, and cone nanostructured perovskite solar cells are considered, and photovoltaic performance is investigated under omnidirectional incidence. The moth-eye configuration has led to the best performance, in which short-circuit current density is increased by 8.4% at normal incidence and by 36.4% at 60° incidence compared to the planar reference [34]. The core-shell geometry provides more degrees of freedom for performance manipulation [35]. In this regard, core-shell spheroidal nanoparticles with metal core and modeled oxide shell (**Figure 4b**), spherical metal-insulator nanoparticles, and metal-metal core-shell nano-cube are integrated, respectively, into the thin film, dye-sensitized, and organic solar cells and plasmon-enhanced light absorption, photocurrent, and efficiency improvement is observed [32, 36].

3.2 Absorption bandwidth enhancement

To widen the bandwidth of the resonant absorbers, the design of multilayered structures and the combined use of elements with different dimensions to excite

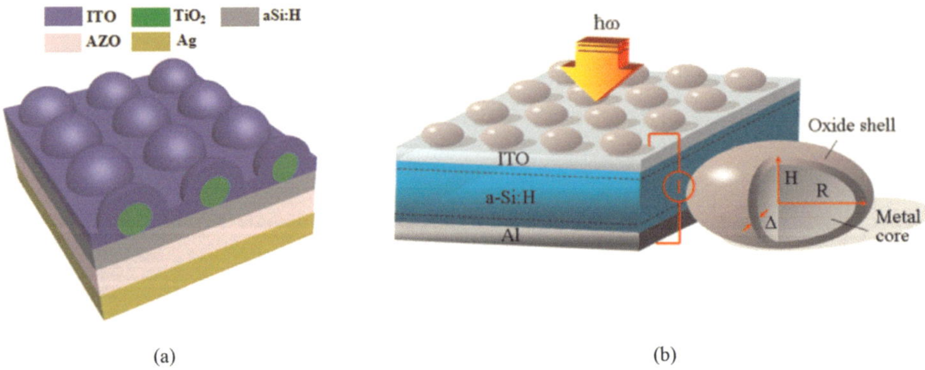

Figure 4.
Thin film solar cell absorption efficiency improvement using (a) dielectric nanoparticles [33] and core-shell plasmonic particles [32].

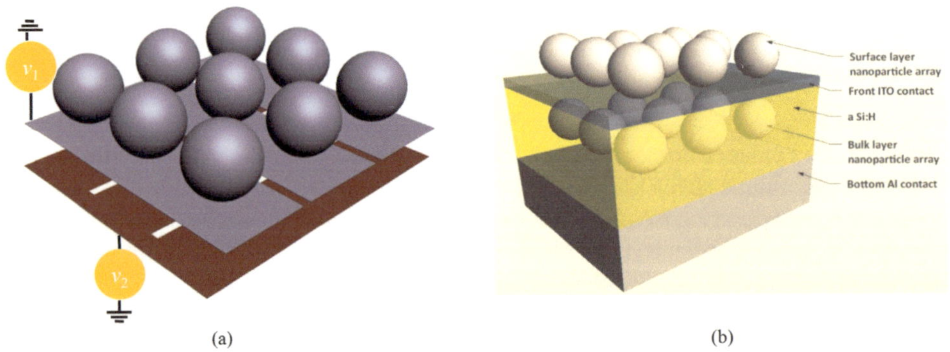

Figure 5.
(a) Absorption bandwidth enhancement using different types of resonances [41] and (b) multiple resonance configuration achieved using multilayered structure [22].

multiple resonances are proposed [37–40]. In the device shown in **Figure 5a**, the hyperbolic nature of the dispersion band supported by the densely packed graphene strips, along with propagating and localized surface plasmon resonances, respectively, provided by the gap plasmons and plasmonic spherical particles, are effectively used to enhance the absorption bandwidth of the multilayer structures [41, 42]. Considering this idea, **Figure 5b** shows a thin film amorphous silicon (a-Si) photovoltaic solar cell in which silver nanoparticle arrays are embedded in the surface and active layer, respectively, being responsible for absorption enhancement in the lower and higher wavelengths [22]. In another design, the rear of a thin film crystalline silicon solar cell is decorated with two layers of silver nanoparticles with different dimensions. By optimizing the radii of the particles, a 9.97% and 9.94% increase, respectively, in short-circuit current density and intergraded quantum efficiency is observed in comparison with the same geometry formed by uniform nanoparticles [43]. Also, randomly distributed metallic nanoparticles with optimized filling factors, laying in the photoactive layer of the thin film solar cell, result in the transportation and localization of light in a broad spectrum [44]. Assuming dielectric particles, multilayered silicon nanoparticles with submicron dimensions are stacked in the ultra-thin photovoltaic solar cell for tailoring the absorption efficiency with Mie scatterers [45].

To illustrate the use of multidimensional elements for the bandwidth enhancement, **Figure 6a** exhibits the wideband absorber design using an oligomer constructed by cylindrical elements. The absorption spectrum, shown in **Figure 6b**, confirms the presence of multiple resonances, originating from the multidimensional elements. The device performs based on transmission elimination using a reflector and reflection elimination arising from the wideband impedance matching with the free space intrinsic impedance. The real and imaginary parts of the retrieved complex normalized surface impedance are extracted via [46–48]:

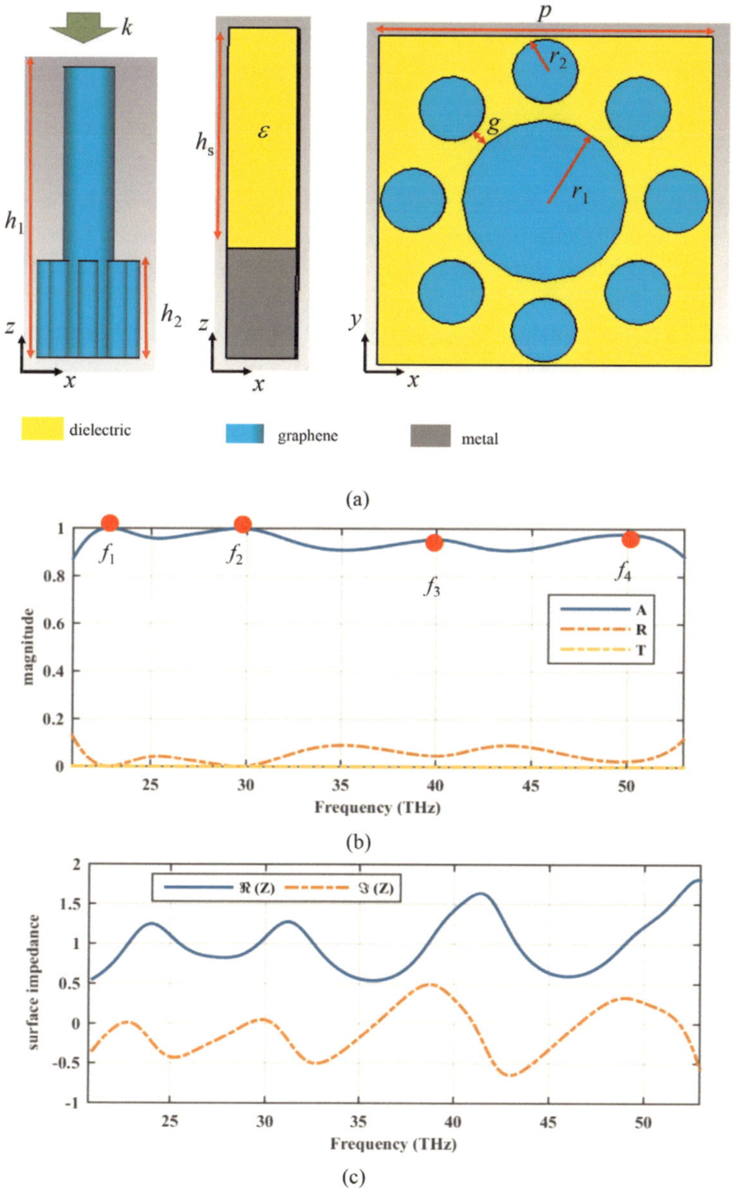

Figure 6.
(a) Absorption bandwidth enhancement using geometrical parameter manipulation and (b)-(c), respectively, absorption spectrum and complex surface impedance of the designed surface [40].

$$Z = \pm\sqrt{\frac{(1+S_{11})^2 - S_{21}^2}{(1-S_{11})^2 - S_{21}^2}} = \pm\frac{1+S_{11}}{1-S_{11}} \tag{3}$$

and they are illustrated in **Figure 6c**. The real and imaginary parts are, respectively, around 1 and 0, confirming the anti-reflection nature of the designed surface. As an instance of the method in the solar cell design, a multiscale metallic fractal nano-carpet configuration is integrated inside the silicon layer of a thin film solar cell for absorption enhancement. The operation of the designed geometry is based on surface plasmon polaritons and localized surface plasmons at different wavelengths, leading to the short circuit current enhancement by a factor of 2.4 [49, 50].

4. Material selection for the solar cell design

Choosing the proper materials for solar cell design is of particular importance. Most common thin film solar cells use Cu (In, Ga) Se2- and CdTe-based photovoltaic technologies, respectively, with module level efficiencies of 19.2% and 18.6%. The high price of In and Ga and the toxicity of Cd is the barrier to the large-scale usage of these solar cells. Metal sulfides and selenides are earth-abundant, highly efficient, environmentally friendly, stable, and cost-effective alternative materials [51]. Also, considering the photothermal absorbers, the absorbers are mainly designed with Au (~1063°C melting point), Ag (~961°C melting point), and Cu metals. The limited bandwidth, disability to tolerate high temperatures, generated in the solar cells with high conversion efficiency, high cost, and oxidation in contact with moisture are the limitations of these materials for developing solar cells [32, 52].

Refractory metals such as titanium (Ti), Tungsten (W) (~3422°C melting point), Chromium (Cr) (~1857°C melting point), Nickel (Ni) (~1453°C melting point), and associated nitrates such as titanium nitride (2930°C melting point) have recently been considered in the design of solar cells due to their high heat tolerance as a result of their high melting point. These corrosion-resistance materials have large imaginary dielectric constants, resulting in a high absorption rate. Moreover, the real part of their dielectric constants is negative, indicating their ability to support surface plasmon resonances [24, 48, 53–56]. In this regard, an all-titanium pyramidal solar cell with a photothermal conversion efficiency of 95.88% in the entire solar spectrum at a temperature of 700°C is designed [57].

Manganese, a transition metal, has the closest constitutive parameters to the ideal metal [58]. **Figure 7a** shows the geometry of truncated pyramidal solar cell design using all-manganese nano-shells [2]. To investigate the absorption capability of manganese material in the solar cell design, the attenuation constant of the wave illuminating the manganese slab is calculated via [59]:

$$\alpha = \frac{\sqrt{2}\pi f}{c}\sqrt{-\varepsilon' + \sqrt{\varepsilon'^2 + \varepsilon''^2}} \tag{4}$$

where prime and double prime show the real and imaginary parts of the dielectric constant, respectively. The complex permittivity of the manganese is illustrated in **Figure 7b**. Also, f is the operating frequency and c is the speed of light in a vacuum. Apart from the influence of the material selection in the absorption efficiency, the use of metallic reflectors for transmission blockage and fulfilling the gradual impedance

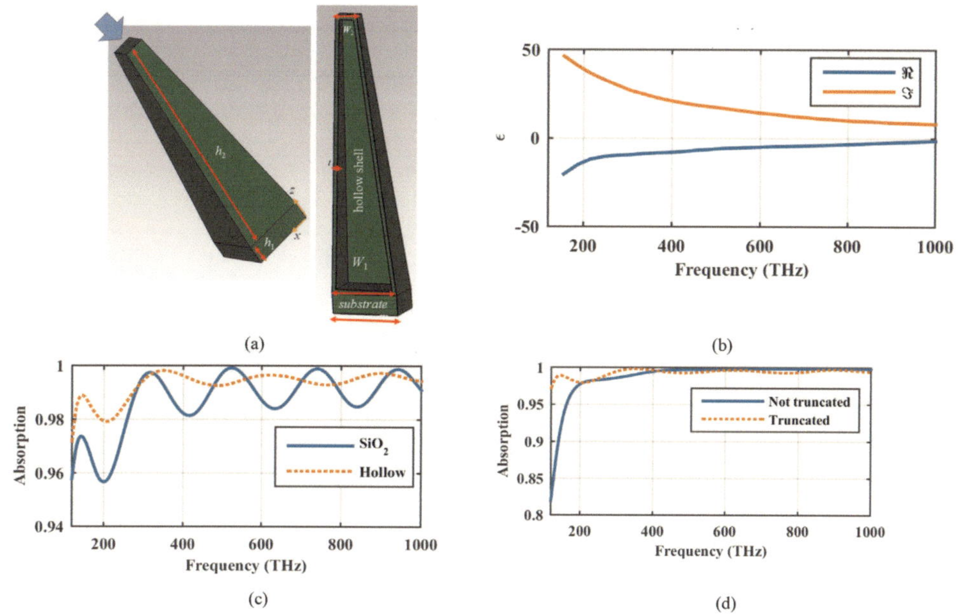

Figure 7.
(a) The geometry of the all-manganese solar cell and (b) the dielectric constant of the manganese in the solar spectrum. (c and d) Respectively show the impact of core material and pyramid tip on the absorption rate [2].

matching by the pyramidal geometry are the main operation mechanisms. Two other design key points for such a wideband performance is the use of hollow and truncated elements for low-frequency absorption enhancement, as shown in **Figure 7c** and **d** [60, 61].

Finally, the use of two-dimensional graphene material in the photothermal solar cell design is revealed. Carbon-based solar cells are recognized for their broadband absorption nature, excellent chemical stability, high thermal stability, and excellent thermal conductivity [62, 63]. Similarly, graphene sheet has high mobility, large optical transparency, excellent mechanical stability, and chemical inertness. It has been used in the design of different types of photovoltaic solar cells such as dye-sensitized solar cells and quantum dot-sensitized solar cells [64]. To evaluate the absorption capability of the graphene in the solar spectrum for potential use in the photothermal solar cell design, the attenuation constant of the illuminating wave to a graphene slab is studied. The attenuation constant (Neper/meter) is related to the real and imaginary parts of the dispersive material permittivity (ε) and permeability (μ) of graphene using [65]:

$$\alpha = \omega\sqrt{\varepsilon_0\mu_0}\left(a^2+b^2\right)^{1/4}\sin\left(\frac{1}{2}\tan^{-1}\left(\frac{a}{b}\right)\right) \quad (5)$$

where prime and double prime respectively denote the real and imaginary parts. The parameters a and b in (5) are defined as: $a = \left(\varepsilon_r'\mu_r' - \varepsilon_r''\mu_r''\right)$ and $b = \left(\varepsilon_r'\mu_r'' + \varepsilon_r''\mu_r'\right)$. As **Figure 8** confirms, the graphene material has a large attenuation constant in the solar spectrum, making it suitable for the absorber design.

Table 1 shows the comparison of graphene's thermal conductivity, density, and specific heat with different metals for the same volume and thermal energy. As can be seen, the thermal conductivity of suspended graphene and graphene on the substrate

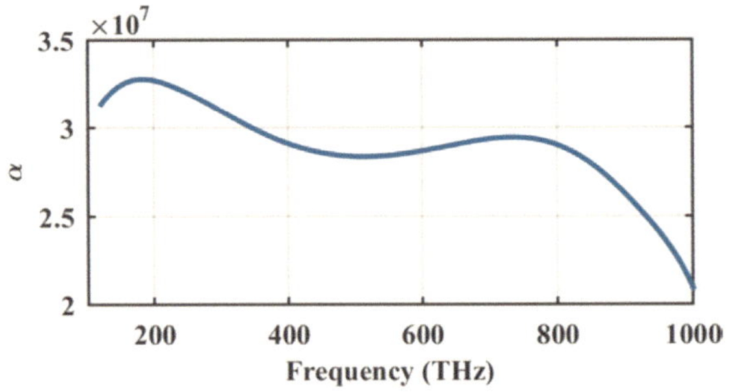

Figure 8.
The attenuation constant of the graphene sheet in the solar spectrum [1].

Material	Thermal conductivity (W m^{-1} K^{-1})	Density (g cm^{-3})	Specific heat (J kg^{-1} K^{-1})	$T_{Metal}/T_{Graphene}$
Al	247	2.7	897	0.6503
Au	317.9	19.3	129	0.6326
Ag	428	10.5	235	0.6383
Zn	113	7.1	338	0.5685
Ni	82.9	8.9	444	0.3986
Cu	401	8.96	385	0.4566
Graphene	2000–4000 (suspend) ~600 (on substrate)	2.25	~700	1

Table 1.
Comparison of graphene's thermal conductivity, density, and specific heat with different metals [63].

is 9.3–48.3 and 1.4–7.2 times of that of metals, respectively. As a result, the transfer of heat absorbed by the graphene-based solar cells is more effective than the metallic samples [63]. Thus, graphene material is a promising candidate for solar cell design.

Figure 9a shows a highly efficient full-spectrum graphene-based solar cell, designed using hollow nanopillars on top of the titanium nitride (TiN) refractory metal substrate. This material combination guarantees the structure's performance in different weather conditions and at high temperatures. The gradual impedance matching of the surface elements with the free space impedance results in a negligible reflection from the surface, and the metallic substrate ensures the elimination of passing waves. Thus, the incoming wave is efficiently trapped in the device, as shown in **Figure 9b**. Note that the use of the truncated cone improves the low-frequency performance of the device and aids in covering the whole solar spectrum. The device has a robust performance for the incoming wave with a wide range of incident angles, as confirmed by **Figure 9c** and **d** [1].

5. Conclusion

Photovoltaic and photothermal systems are considered the two main solar cell design technologies, and their design key points are introduced in this chapter. The efficiency and the operating bandwidth are important factors for evaluating the performance of solar cells. To reach efficient solar cells, it is required to optimize the surface geometry in terms of shape, material, and the number of layers. In general,

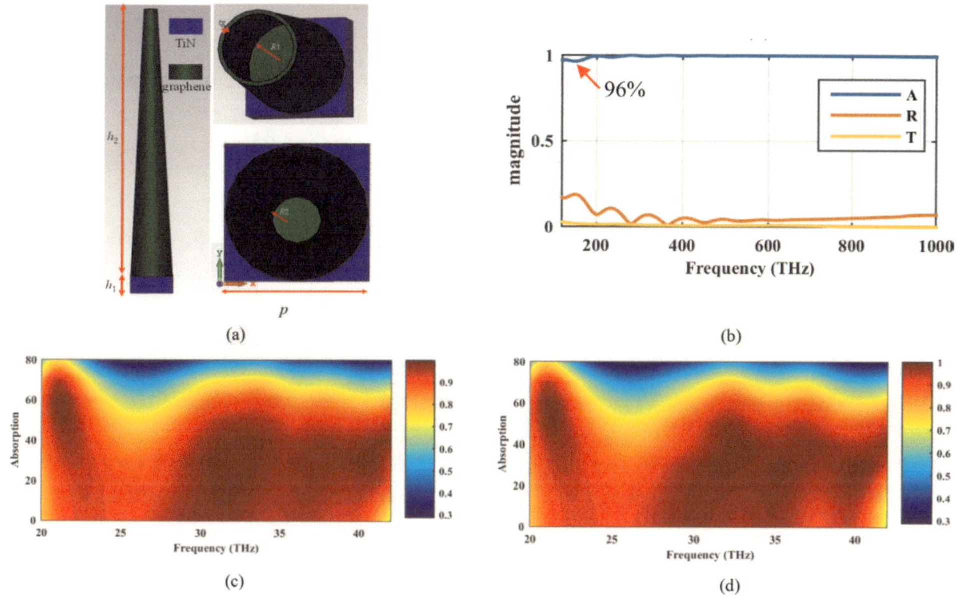

Figure 9.
Entire solar spectrum absorption coverage with (a) graphene-based nano-pillar absorber and (b) its reflection, transmission, and absorption spectra. (c and d) The sensitivity of the absorption rate to the incident angle respectively for the TE and TM waves [1].

dielectric materials lead to better performance in comparison to noble metals. To design broadband absorbers, multilayered or multiresonance configurations are proposed. Alternatively, pyramidal/conical geometries, supporting gradual impedance matching with the free space intrinsic impedance, can be exploited. In these geometries, the tip truncation and use of hollow cores improve the low-frequency efficiency. Moreover, environmentally friendly and heat-tolerant materials are the best choice for practical solar cell designs and different types of two-dimensional materials can also be beneficial for solar cell performance optimization. Considering all the factors, novel solar cells with improved performance can be proposed for future applications.

Author details

Shiva Hayati Raad* and Zahra Atlasbaf
Department of Electrical and Computer Engineering, Tarbiat Modares University, Tehran, Iran

*Address all correspondence to: shiva.hayati@modares.ac.ir

References

[1] Raad SH, Atlasbaf Z. Solar cell design using graphene-based hollow nano-pillars. Scientific Reports. 2021;**11**(1):1-8

[2] Raad SH, Atlasbaf Z. Full coverage of the solar spectrum and beyond using all-manganese plasmonic shell array. Plasmonics. 2022;**17**(2):851-857

[3] David MM, Timonah S. Solar solutions for the future. In: Poorva S, Ashwini K, editors. Recent Advances in Multifunctional Perovskite Materials. Rijeka: IntechOpen; 2022. p. 9

[4] Li X et al. High-performance layer-by-layer organic solar cells enabled by non-halogenated solvent with 17.89% efficiency. Chemical Engineering Journal. 2023;**452**:139496

[5] Yella A et al. Porphyrin-sensitized solar cells with cobalt (II/III)-based redox electrolyte exceed 12 percent efficiency. Science. 2011;**334**(6056): 629-634. DOI: 10.1126/science.1209688

[6] Hao M et al. Ligand-assisted cation-exchange engineering for high-efficiency colloidal $Cs_{1-x}FA_xPbI_3$ quantum dot solar cells with reduced phase segregation. Nature Energy. 2020; **5**(1):79-88. DOI: 10.1038/s41560-019-0535-7

[7] Olaimat MM, Yousefi L, Ramahi OM. Using plasmonics and nanoparticles to enhance the efficiency of solar cells: A review of latest technologies. JOSA B. 2021;**38**(2):638-651

[8] Chen X, Jia B, Zhang Y, Gu M. Exceeding the limit of plasmonic light trapping in textured screen-printed solar cells using Al nanoparticles and wrinkle-like graphene sheets. Light: Science & Applications. 2013;**2**(8): e92-e92

[9] Tsai M-L et al. Monolayer MoS2 heterojunction solar cells. ACS Nano. 2014;**8**(8):8317-8322. DOI: 10.1021/nn502776h

[10] Makableh YF, Al-Fandi M, Khasawneh M, Tavares CJ. Comprehensive design analysis of ZnO anti-reflection nanostructures for Si solar cells. Superlattices and Microstructures. 2018;**124**:1-9

[11] Raad SH. Analysis and design of absorbers for electromagnetic compatibility applications. In: Recent Topics in Electromagnetic Compatibility. London, UK: IntechOpen; 2021

[12] Chen Z. Common RF absorbers evaluations in W-band (75-100 GHz). In: 2017 IEEE International Symposium on Electromagnetic Compatibility & Signal/Power Integrity (EMCSI). United States. 2017. pp. 1-31. DOI: 10.1109/ISEMC.2017.8078025

[13] Baker-Finch SC, McIntosh KR. Reflection of normally incident light from silicon solar cells with pyramidal texture. Progress in Photovoltaics: Research and Applications. 2011;**19**(4):406-416. DOI: 10.1002/pip.1050

[14] Liang H, Wang F, Yang L, Cheng Z, Shuai Y, Tan H. Progress in full spectrum solar energy utilization by spectral beam splitting hybrid PV/T system. Renewable Sustainable Energy Reviews. 2021;**141**: 110785

[15] Hamdy H, Abdel-Latif GY, El-Agamy M, El-Mikati H, Hameed MFO, Obayya S. Wavelength-selective metamaterial absorber based on 2D split rhombus grating for thermophotovoltaic solar cell. Optical and Quantum Electronics. 2022;**54**(2):1-12

[16] Aliyariyan M, Fathi D, Eskandari M, Mohammadi MH. Simulation and investigation of perovskite/nano-pyramidal GeSe solar cell: Realizing high efficiency by controllable light trapping. Solar Energy. 2021;**214**:310-318

[17] Sutter J et al. Tailored nanostructures for light management in silicon heterojunction solar cells. Solar RRL. 2020;**4**(12):2000484

[18] Green M, Dunlop E, Hohl-Ebinger J, Yoshita M, Kopidakis N, Hao X. Solar cell efficiency tables (version 57). Progress in Photovoltaics: Research and Applications. 2021;**29**(1):3-15

[19] Atwater HA, Polman A. Plasmonics for improved photovoltaic devices. Nature Materials. 2010;**9**(3):205-213

[20] Dal Negro L, Boriskina SV. Deterministic aperiodic nanostructures for photonics and plasmonics applications. Laser & Photonics Reviews. 2012;**6**(2):178-218

[21] Ma C et al. Plasmonic-enhanced light harvesting and perovskite solar cell performance using Au biometric dimers with broadband structural darkness. Solar RRL. 2019;**3**(8):1900138

[22] Krishnan A, Das S, Krishna SR, Khan MZA. Multilayer nanoparticle arrays for broad-spectrum absorption enhancement in thin film solar cells. Optics Express. 2014;**22**(103):A800-A811

[23] Yu P et al. A numerical research of wideband solar absorber based on refractory metal from visible to near-infrared. Optical Materials. 2019;**97**:109400

[24] Liang Q, Duan H, Zhu X, Chen X, Xia X. Solar thermal absorber based on dielectric-filled two-dimensional nickel grating. Optical Materials Express. 2019;**9**(8):3193-3203

[25] Kumar R, Singh BK, Pandey PC. Cone-Shaped Resonator-Based Highly Efficient Broadband Polarization-Independent Metamaterial Absorber for Solar Energy Harvesting. 2022. PREPRINT (Version 1) available at Research Square. DOI: 10.21203/rs.3.rs-2280426/v1

[26] Wang DH et al. Enhancement of donor-acceptor polymer bulk heterojunction solar cell power conversion efficiencies by addition of Au nanoparticles. Angewandte Chemie. 2011;**123**(24):5633-5637

[27] Kozanoglu D, Apaydin DH, Cirpan A, Esenturk EN. Power conversion efficiency enhancement of organic solar cells by addition of gold nanostars, nanorods, and nanospheres. Organic Electronics. 2013;**14**(7):1720-1727

[28] Wang W et al. Efficient light trapping in organic solar cell using a short-pitched hexagonal array of metallic nanocylinders. IEEE Photonics Journal. 2016;**8**(5):1-9. DOI: 10.1109/JPHOT.2016.2614601

[29] Raad SH, Atlasbaf Z. Broadband/multiband absorption through surface plasmon engineering in graphene-wrapped nanospheres. Applied Optics. 2020;**59**(28):8909-8917

[30] Hajjiah A, Kandas I, Shehata N. Efficiency enhancement of perovskite solar cells with plasmonic nanoparticles: A simulation study. Materials. 2018;**11**(9):1626

[31] Akimov YA, Koh W, Sian S, Ren S. Nanoparticle-enhanced thin film solar cells: Metallic or dielectric nanoparticles? Applied Physics Letters. 2010;**96**(7):073111

[32] Akimov YA, Koh WS. Design of plasmonic nanoparticles for efficient subwavelength light trapping in thin-film solar cells. Plasmonics. 2011;**6**(1): 155-161

[33] Yang Z, Gao P, Zhang C, Li X, Ye J. Scattering effect of the high-index dielectric nanospheres for high performance hydrogenated amorphous silicon thin-film solar cells. Scientific Reports. 2016;**6**(1):1-7

[34] Du D et al. The broadband and omnidirectional antireflective performance of perovskite solar cells with curved nanostructures. Solar Energy. 2021;**224**:10-17

[35] Raad SH, Atlasbaf Z. Tunable optical absorption using graphene covered core-shell nano-spheres. In: Iranian Conference on Electrical Engineering (ICEE). IEEE; 2018. pp. 98-102

[36] Brown MD et al. Plasmonic dye-sensitized solar cells using core-shell metal-insulator nanoparticles. Nano Letters. 2011;**11**(2):438-445. DOI: 10.1021/nl1031106

[37] Sun L, Wang X, Yu Z, Huang J, Deng L. Patterned AlN ceramic for high-temperature broadband reflection reduction. Journal of Physics D: Applied Physics. 2019;**52**(23):235102

[38] Wang H, Wang L. Perfect selective metamaterial solar absorbers. Optics Express. 2013;**21**(106):A1078-A1093

[39] Zhang Z et al. Broadband metamaterial absorber for low-frequency microwave absorption in the S-band and C-band. Journal of Magnetism Magnetic Materials. 2020;**497**:166075

[40] Raad SH, Atlasbaf Z. Broadband continuous/discrete spectrum optical absorber using graphene-wrapped fractal oligomers. Optics Express. 2020; **28**(12):18049-18058

[41] Raad SH, Atlasbaf Z, Zapata-Rodríguez CJ. Broadband absorption using all-graphene grating-coupled nanoparticles on a reflector. Scientific Reports. 2020;**10**(1):1-15

[42] Raad SH, Atlasbaf Z, Shahabadi M, Rashed-Mohassel J. Dyadic Green's function for the tensor surface conductivity boundary condition. IEEE Transactions on Magnetics. 2019;**55**(11):1-7

[43] Zhang S et al. Absorption enhancement in thin film solar cells with bilayer silver nanoparticle arrays. Journal of Physics Communications. 2018;**2**(5): 055032

[44] Piralaee M, Asgari A. Modeling of optimum light absorption in a random plasmonic solar cell using effective medium theory. Optical Materials. 2016; **62**:399-402

[45] Mirnaziry SR, Shameli MA, Yousefi L. Design and analysis of multi-layer silicon nanoparticle solar cells. Scientific Reports. 2022;**12**(1):1-13

[46] Xiong H, Ji Q, Bashir T, Yang F. Dual-controlled broadband terahertz absorber based on graphene and Dirac semimetal. Optics Express. 2020;**28**(9): 13884-13894

[47] Lai S, Wu Y, Zhu X, Gu W, Wu W. An optically transparent ultra-broadband microwave absorber. IEEE Photonics Journal. 2017;**9**(6):1-10

[48] Huang Z, Wang B. Ultra-broadband metamaterial absorber for capturing solar energy from visible to near-infrared. Surfaces and Interfaces. 2022;**33**:102244

[49] Shameli MA, Yousefi L. Absorption enhanced thin-film solar cells using

fractal nanostructures. IET Optoelectronics. 2021;**15**(5):248-253

[50] Kazerooni H, Khavasi A. Plasmonic fractals: Ultra-broadband light trapping in thin film solar cells by a Sierpinski nano-carpet. Optical and Quantum Electronics. 2014;**46**(6):751-757

[51] Mavlonov A et al. A review of Sb2Se3 photovoltaic absorber materials and thin-film solar cells. Solar Energy. 2020; **201**:227-246. DOI: 10.1016/j.solener.2020.03.009

[52] Li W et al. Refractory plasmonics with titanium nitride: Broadband metamaterial absorber. Advanced Materials. 2014;**26**(47):7959-7965. DOI: 10.1002/adma.201401874

[53] Kim I, So S, Rana AS, Mehmood MQ, Rho J. Thermally robust ring-shaped chromium perfect absorber of visible light. Nano. 2018;**7**(11):1827-1833

[54] Yu P et al. Ultra-wideband solar absorber based on refractory titanium metal. Renewable Energy. 2020;**158**: 227-235

[55] Li H, Niu J, Zhang C, Niu G, Ye X, Xie C. Ultra-broadband high-efficiency solar absorber based on double-size cross-shaped refractory metals. Nanomaterials. 2020;**10**(3):1-11. DOI: 10.3390/nano10030552

[56] Song D, Zhang K, Qian M, Liu Y, Wu X, Yu K. Ultra-broadband perfect absorber based on titanium nanoarrays for harvesting solar energy. Nanomaterials. 2023;**13**(1):91. [Online]. Available: https://www.mdpi.com/2079-4991/13/1/91

[57] Qi B, Chen W, Niu T, Mei Z. Ultra-broadband refractory all-metal metamaterial selective absorber for solar thermal energy conversion. Nanomaterials. 2021;**11**(8):1-10. DOI: 10.3390/nano11081872

[58] Alizadeh M, Khavasi A, Butun B, Ozbay E. Large-area, cost-effective, ultra-broadband perfect absorber utilizing manganese in metal-insulator-metal structure. Scientific Reports. 2018; **8**(1):1-13

[59] Derakhshani M, Taheri-Nassaj E, Jazirehpour M, Masoudpanah S. Structural, magnetic, and gigahertz-range electromagnetic wave absorption properties of bulk Ni–Zn ferrite. Scientific Reports. 2021;**11**(1):1-13

[60] Ding F, Jin Y, Li B, Cheng H, Mo L, He S. Ultrabroadband strong light absorption based on thin multilayered metamaterials. Laser Photonics Reviews. 2014;**8**(6):946-953

[61] Dang PT et al. Efficient broadband truncated-pyramid-based metamaterial absorber in the visible and near-infrared regions. Crystals. 2020;**10**(9):784

[62] Ren H et al. Hierarchical graphene foam for efficient omnidirectional solar–thermal energy conversion. Advanced Materials. 2017;**29**(38):1702590. DOI: 10.1002/adma.201702590

[63] Lin K-T, Lin H, Yang T, Jia B. Structured graphene metamaterial selective absorbers for high efficiency and omnidirectional solar thermal energy conversion. Nature Communications. 2020;**11**(1):1-10

[64] Das S, Sudhagar P, Kang YS, Choi W. Graphene synthesis and application for solar cells. Journal of Materials Research. 2014;**29**(3):299-319

[65] Xu Q, Huang Y. Anechoic and Reverberation Chambers: Theory, Design, and Measurements. Wiley-IEEE Press; 2019

Chapter 4

Operation of Photovoltaic Panels in Stand-alone Applications

Ali Lamkaddem, Hajar Chadli, Khalid Salmi, Rachid Malek, Olivier Deblecker, Khalil Kassmi and Najib Bachiri

Abstract

In this chapter, we propose the analysis of the maximum power point (MPP) of photovoltaic panels (PV) in a renewable energy application. From the current–voltage characteristics, we deduced the MPP of a PV panel and specified the use of a power block (DC/DC converter) controlled by an MPPT control. In the case of an MPPT control of type perturb and observe, we realized the photovoltaic system that heats a photovoltaic solar cooker, taking into account this MPPT command. The experimentation of this application, during a sunny day, shows that the MPPT control carries out its role correctly, such as optimal operation of the PV panels and heating of the cooker by the maximum power supplied by the PV panels. The analysis of all the results shows an excellent agreement between the experiment and the simulation of the operation of the photovoltaic system which made it possible to operate the photovoltaic panels around their MPP, over the course of the sun. Under these conditions, the efficiency of the proposed DC/DC converter, with a power of 500 Wp, is of the order of 97%.

Keywords: energy, photovoltaic panels, maximum power point, MPPT command, DC/DC converter, efficiency

1. Introduction

Solar photovoltaic is the most widely used renewable energy source with relatively high accessibility in many parts of the world [1–4]. In recent years, the development of the solar PV market and the use of this technology worldwide has been increasing at annual rates of 35–40%. This rapid expansion has developed rapidly due to a sharp decline in PV prices and increased attention to the importance of sustainable energy [4–8].

In addition, the market development of solar photovoltaic (PV) applications is increasing, due to their contributions to environmental protection, and use as an alternative sustainable solution to the energy crisis. In the short and long term, PV technology is considered the main source of electricity generation in various applications such as [9–16] solar desalination, solar cookers, solar cooling, and air conditioning. In PV technology, the problem that arises is the operation of PV panels around their maximum power points (MPP). Regardless of the applications used, stand-alone or grid-connected, the design of a power matching system (DC/DC converter) of the PV panels with its MPPT control (maximum power point tracking) [17, 18] is necessary to produce the maximum electrical energy during the operation of these applications [19]. In the literature, several MPPT controls are proposed [20, 21]. Each method has its advantages and disadvantages. Currently, in the context of PV energy applications that we are developing in the framework of national and international projects, we use the perturb and observe control [21–24], due to its ease of implementation and accuracy of convergence to the maximum power point of the PV panels.

In this chapter, we analyze the maximum power point (MPP) of PV panels and the PV systems that maximize this power during the operation of an application. Particular attention is paid to the description of the MPPT control, the perturb and observe type, and its application in PV solar cookers, with a power of 600 Wp.

2. Optimization of the operation of photovoltaic (PV) panels

2.1 Electrical model of a PV panel cell

Figure 1 shows the equivalent electrical circuit of a cell (Junction P/N) of a PV panel, illuminated by radiation of intensity "Le" [25]. This circuit is formed by a diode (D), a current source (I_{CC}), a contact series resistance (R_s), and a parallel resistance (R_{sh}), which represents the leakage current of the cell. From this diagram, the current of the PV cell (I) is expressed as a function of the voltage V according to the expression [25–29]:

$$I = I_{CC} - I_s \left\{ \exp\left[\frac{q(V + IR_s)}{A \cdot K \cdot T}\right] - 1 \right\} - \left(\frac{V + IR_s}{R_{sh}}\right) \tag{1}$$

where
I_s: Reverse saturation current of the diode (A),
q: Electric charge of an electron = 1602 × 10^{-19} Coulomb,
A: The diode ideality factor,
K: Boltzmann constant 1.381 × 10^{-23} J/K,
T: Cell temperature (°C).

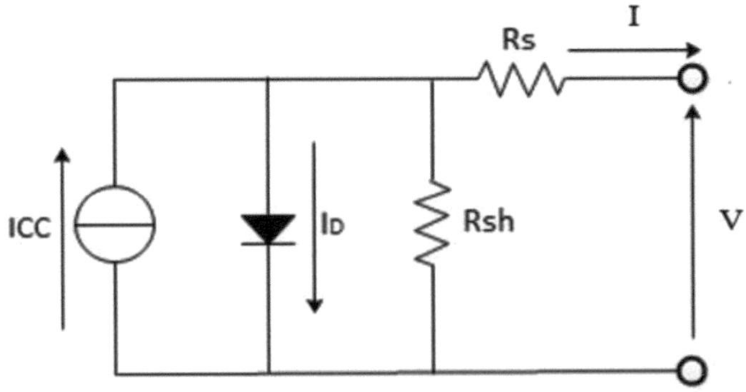

Figure 1.
The diagram of the electrical circuit of a cell of a PV panel.

In this expression, the current I_{CC} depends very little on the temperature and varies with the intensity of the illumination according to the equation [30, 31]:

$$I_{CC} = A.Le + B \qquad (2)$$

where

A and B are two constant factors [30, 31].

As part of our application, we used PV panels that have cells, where A = 0.00932; B = 0.0055; very low contact resistance and very high parallel resistance. Under these conditions, the current I is written as a function of the voltage V according to the expression:

$$I = A.Le + B - I_s\left\{\exp\left[\frac{q(V)}{A \cdot K \cdot T}\right] - 1\right\} \qquad (3)$$

2.2 Electrical characteristics of PV panels

As part of our experiment, we used PV panels (**Figure 2**), 300 Wp, formed by $N_s = 80$ identical cells in series. The expression of the current of the I_{PV} panels is written according to the intensity of the illumination (L_e) and the voltage V_{PV} according to the equations:

$$I_{PV} = 0.00932 * L_e + 0.0055 - I_s\left\{\exp\left[\frac{q(V_{PV})}{80.A \cdot K \cdot T}\right] - 1\right\} \qquad (4)$$

The typical experimental electrical characteristics (current–voltage and power—voltage) of the PV panels (**Figure 2**), each with a power of 300 Wp, for illuminations ranging from 300 W/m^2 to 900 W/m^2, are shown in **Figure 3**. From these characteristics, we have determined and represented in **Table 1** the optimal electrical values:

Photovoltaic Panels
(Type 1) (305 W)

Figure 2.
Photos of 600 Wp photovoltaic panels (Type 1) installed in the laboratory.

	Le (W/m^2)	Vopt (V)	Iopt (A)	Ropt (Ω)	Popt (W)
Panel PV (305 W):Type 1	300	25.4	2.87	8.85	72.9
	500	25.5	4.78	5.33	121.9
	900	25.4	8.61	2.95	218.7

Table 1.
Optimal electrical values of a PV panel, from ***Figure** 2, as a function of illuminations. Temperature = 25°C.*

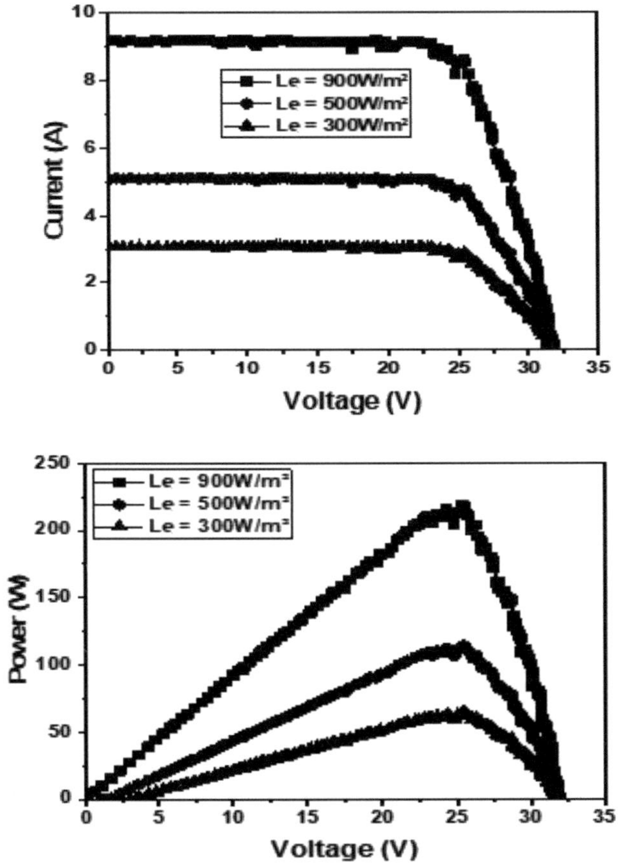

Figure 3.
Current–voltage-power characteristics of the photovoltaic panel used (Type 1), for three illuminations (300, 500, and 900 W/m²). Ambient temperature = 25°C.

voltage, current, maximum power point (MPP), and optimal resistance. Therefore, we can deduce, when the illuminance varies from 300 to 900 W/m^2, the optimal power and resistance vary from 72.9 to 218.7 W and from 8.85 to 95 Ω, respectively. In a PV application, to ensure the production of optimal electrical energy and reduce the cost of kWh, the operation of PV panels, over the sun, must follow these variations.

Moreover, in the design of PV system blocks, these optimal values allow for determining the size of each electrical component of the power, and the nature of the MPPT control that optimizes the operation of the PV panels, during the operation of the applications throughout the day.

2.3 MPPT command: perturb and observe

In this section, we describe the basic functioning of the MPPT command type perturb and observe that we used in all the applications developed in the laboratory. To do this, the PV panel adaptation block (**Figure 3**) generates a PWM signal of frequency f and duty cycle α, following the execution of an MPPT algorithm (**Figure 4**), which controls the DC/DC converters. The basic principle of the P&O method consists in disturbing the voltage of the PV panels (Vpv) and observing its impact on the variation of the output power of the PV panel, following the steps:

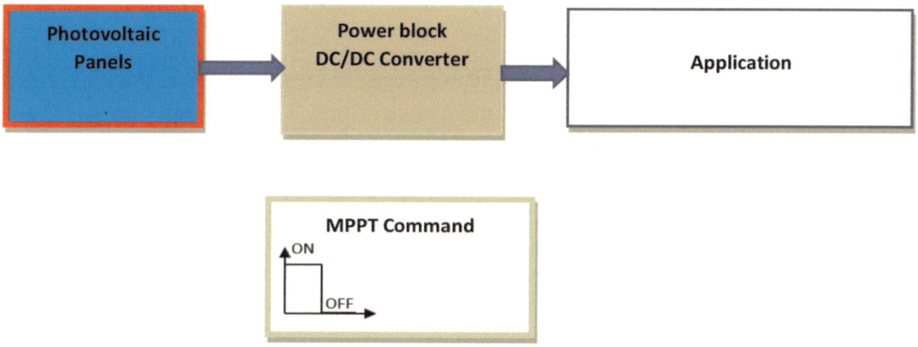

Figure 4.
PV panel adaptation system.

- Acquire and store the electrical quantities of the converter: voltages, currents, powers, efficiency, and duty cycle.
- Once all the quantities have been acquired, the microcontroller generates the PWM signals with a variable duty cycle α.
- At each cycle, Vpv and Ipv are measured to calculate Ppv (k). This value of Ppv (k) is compared to the value of Ppv (k-1) calculated in the previous cycle. When the output power has increased, Vpv is adjusted in the same direction as in the previous cycle. If the output power has decreased, Vpv is adjusted in the opposite direction as in the previous cycle. When the maximum power point is reached, this means that Vpv is around the optimal value.

3. Application: Photovoltaic solar cooker

In this section, we have analyzed the optimization of the PV panels by the system of **Figure 4** powered by two PV panels of **Figure 2** in series. The PV application consists of heating a cooker with 600 Wp PV energy. In the following sections, we present the model and the optimization of the PV panels used, then the operation and the validation of the PV system of **Figure 4** in the case of a cooker with PV energy.

3.1 Model of PV panels

In the case of our 600 Wp application, the expression of the current Igene as a function of the voltage Vgene of the PV generator, formed by two panels of **Figure 2** in series, is written for a given illumination according to the expression:

$$I_{PV} = 0.00932 * L_e + 0.0055 - I_s\left\{ \exp\left[\frac{q(V_{PV})}{160.A \cdot K \cdot T}\right] - 1\right\} \quad (5)$$

As previously (Section 2.2) mentioned, experimentation with this generator when the intensity of the illumination varies from 300 to 900 W shows optimal voltages, currents, resistances, and powers, which vary from 50.8 to 51.0 V; 2.87 to 8.61 A; 17.7 to 25.8 Ω, and 145.8 to 437.4 W.

3.2 Operation of the cooker

We experimented, during 1 day, the PV system of **Figure 3** in the case of the heating of 1 liter of water of a photovoltaic solar cooker. The principle of operation of this cooker is the supply of thermal resistance by optimal energy supplied by the PV panels, *via* a DC/DC converter, controlled by an MPPT command "perturb and observe" (**Figure 4**). We recorded the meteorological electrical quantities (intensity of illumination and ambient temperature), the electrical quantities (voltage, current, and power) at the input and output of the DC/DC converter, and the heating temperatures of the water and thermal resistance (**Figure 5**). The results obtained show:

- Intensity of illumination and ambient temperature, respectively, reach maximum values of around 998 W/m^2 and 22°C.
- The voltages and currents at the input of the converter are, respectively, around 52.4 V and 9.1 A and at the output around 96.9 V and 4.8 A.
- The powers at the input and there at the output of the converter are, respectively, of the order of 476.8 W and 465.1 W, that is, an efficiency of 97%.
- The temperature of the thermal resistance reaches the value of 300°C after 5 seconds of heating (i.e., 60°C/s), and the maximum value reached 660°C after 20 seconds.

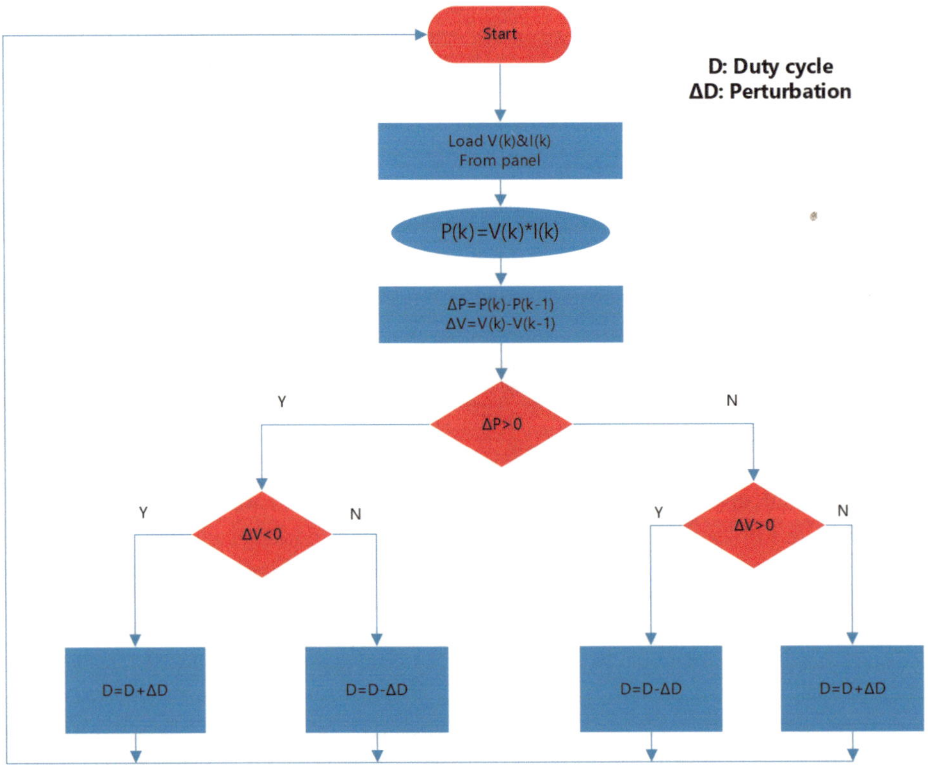

Figure 5.
MPPT command flowchart Perturb and Observe.

- During 5 minutes of heating, the water temperature varies from 24–49°C, that is, 3.8°C/min. In this case, the thermal efficiency is around 75%.
- After 18 minutes of heating, the water reaches the maximum boiling temperature, which is 98°C.

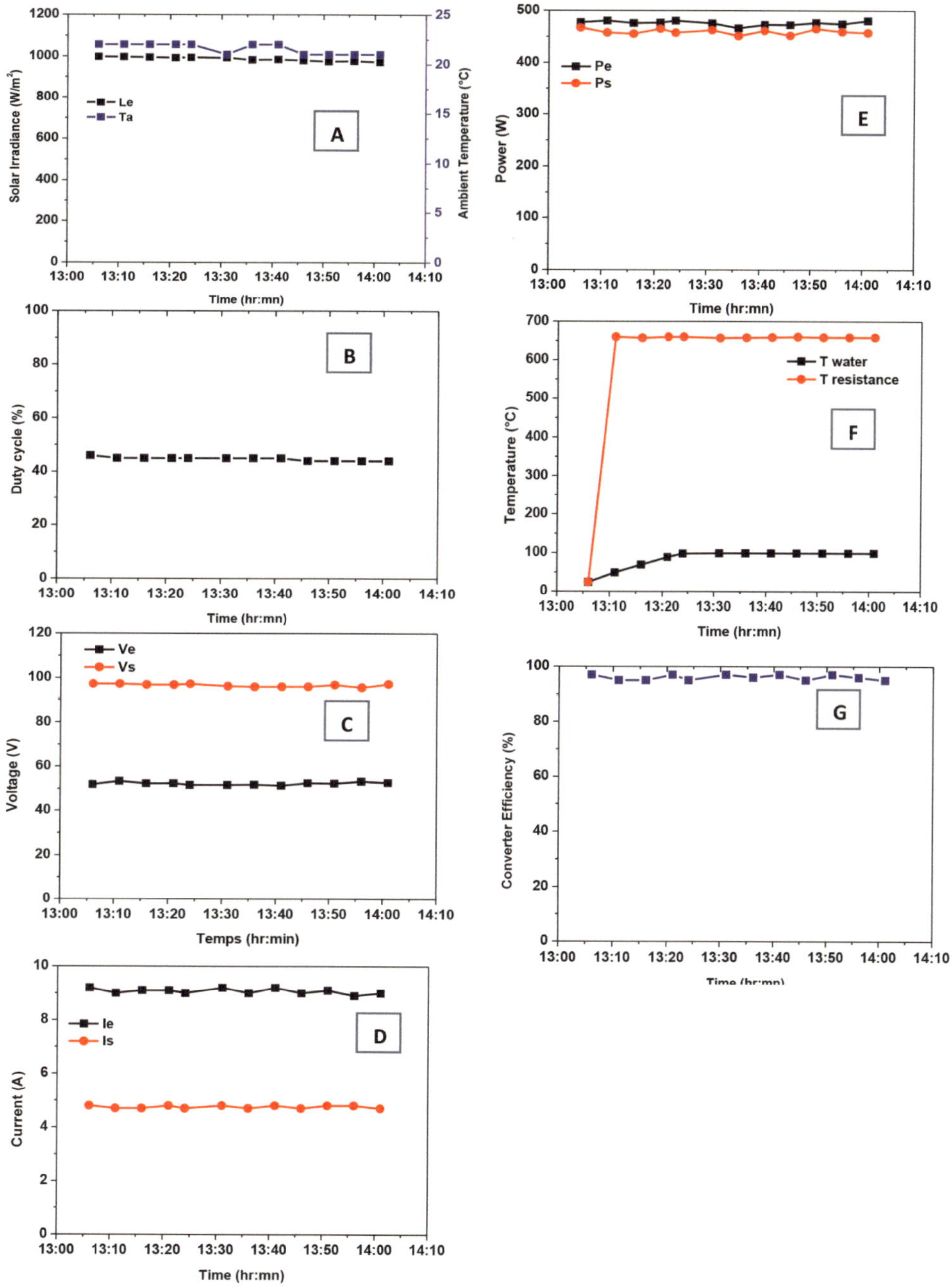

Figure 6.
*Experimentation of the cooker (hot plate) (**Figure 3**) during the heating of 1 liter of water. A: Illumination and external temperature. B: Duty cycle of the DC/DC converter. C: Input and output voltage of the DC/DC converter. D: Input and output current of the DC/DC converter. E: Power of Photovoltaic generator and heating resistor. F: Efficiency of the DC/DC converter. G: Temperature of the heating resistor and water.*

All the results obtained on the PV system, controlled by the MPPT command, show that the proposed MPPT command plays its role correctly, and the solar cooker is heated by the maximum power provided by the PV panels (**Figure 6**).

4. Conclusion

In this chapter, we have analyzed the operation of photovoltaic (PV) panels around their maximum power points (MPP). We recalled the equations and electrical models that govern the operation of these panels and showed that knowledge of their electrical quantities, over the sun, is essential to design the different blocks of the PV system, which control their operation in a PV application. As part of our experiment, when the intensity of the illumination varies from 300 to 900 W, the optimum voltages, currents, powers, and resistances of a PV panel (300 Wp), vary, respectively, from 25.4 to 25, 5 V; 2.87 to 8.61 A; 72.9 to 218.7 W and 8.85 to 2.95 Ω. Then, we showed that these electrical quantities are fixed through a power block (DC/DC converter), controlled by an MPPT command, which generates a variable duty cycle PWM signal. We specified that the most established MPPT command is that of perturb and observe since it presents the stability and the precision of convergence toward the MPP. In the case of a solar photovoltaic application (solar cooker), we have followed the role of this control in the optimization of the operation of the 600 Wp PV panels (two PV panels in series). We have shown that, during a sunny day, where the intensity of the illumination varies from 950 to 1000 W/m^2, the duty cycle of the PWM signal varies from 0.42 to 0.45, and the operation of the PV panels converges toward their optimum electrical quantities (voltage, current, optimum resistance, and power of the order of: 51.17 V, 9.2 A, 5.56 Ω, and 470 W), and the temperature of a liter of water has reached the boiling temperature of 90°C after 23 min of heating. All the results obtained show the role of the power unit and its perturb and observe type command to optimize the operation of PV panels, in a PV application, over the sun.

Acknowledgements

This research is supported within the framework of the projects:

- Morocco-Wallonie Brussels Cooperation Program (2018–2022), Wallonie-Bruxelles-International (Belgium), project 4, n°2.
- Europe-Africa Research and Innovation call on Renewable Energy LEAP-RE, Project No 963530: SoCoNexGen, Solar Indoor Cooking Systems of the Next Generation, 2022–2025.
- National Initiative for Human Development INDH, Berkane Province, Morocco, project 2017/29.

Author details

Ali Lamkaddem[1], Hajar Chadli[1], Khalid Salmi[1], Rachid Malek[1], Olivier Deblecker[2]*, Khalil Kassmi[1]* and Najib Bachiri[3]

1 Faculty of Science, Department of Physics, Laboratory of Electromagnetic, Signal Processing and Renewable Energy LESPRE, Team Electronic Materials and Renewable Energy EMRE, Mohamed First University, Oujda, Morocco

2 Association Humain and Environnement of Berkane (AHEB), Berkane, Morocco

3 Polytech. Mons - Electrical Power Engineering Unit, University of Mons, Mons, Belgium

*Address all correspondence to: olivier.deblecker@umons.ac.be and khkassmi@yahoo.fr; k.kassmi@ump.ac.ma

References

[1] Hernández-Callejo L, Gallardo-Saavedra S, Alonso-Gómez V. A review of photovoltaic systems: Design, operation and maintenance. Solar Energy. 2019;**188**:426-440

[2] Husain AA, Hasan WZW, Shafie S, Hamidon MN, Pandey SS. A review of transparent solar photovoltaic technologies. Renewable and Sustainable Energy Reviews. 2018;**94**:779-791

[3] Al-Shahri OA, Ismail FB, Hannan MA, Lipu MH, Al-Shetwi AQ, Begum RA, et al. Solar photovoltaic energy optimization methods, challenges and issues: A comprehensive review. Journal of Cleaner Production. 2021;**284**: 125465

[4] Khodayar M, Khodayar ME, Jalali SMJ. Deep learning for pattern recognition of photovoltaic energy generation. The Electricity Journal. 2021; **34**(1):106882

[5] Song Z, Liu J, Yang H. Air pollution and soiling implications for solar photovoltaic power generation: A comprehensive review. Applied Energy. 2021;**298**:117247

[6] Sampaio PGV, González MOA. Photovoltaic solar energy: Conceptual framework. Renewable and Sustainable Energy Reviews. 2017;**74**:590-601

[7] Mbungu NT, Naidoo RM, Bansal RC, Siti MW, Tungadio DH. An overview of renewable energy resources and grid integration for commercial building applications. Journal of Energy Storage. 2020;**29**:101385

[8] Hemeida AM, El-Ahmar MH, El-Sayed AM, Hasanien HM, Alkhalaf S, Esmail MFC, et al. Optimum design of hybrid wind/PV energy system for remote area. Ain Shams Engineering Journal. 2020;**11**(1):11-23

[9] Talbi S, Atmane I, Elmoussaoui N, Kassmi K, Deblecker O. Feasibility of a box-type solar cooker powered by photovoltaic energy. In: 2019 7th International Renewable and Sustainable Energy Conference (IRSEC). Agadir, Morocco; IEEE; 2019. pp. 1-4. Available from: https://ieeexplore.ieee.org/abstract/document/9078275; http://www.med-space.org/irsec19/

[10] Atmane I, El Moussaoui N, Kassmi K, Deblecker O, Bachiri N. Development of an innovative cooker (hot plate) with photovoltaic solar energy. Journal of Energy Storage. 2021; **36**:102399

[11] Lamkaddem A, Moussaoui NE, Rhiat M, Malek R, Kassmi K, Deblecker O, et al. System for powering autonomous solar cookers by batteries. Scientific African. 2022;**17**:e01349

[12] Anand B, Shankar R, Murugavelh S, Rivera W, Prasad KM, Nagarajan R. A review on solar photovoltaic thermal integrated desalination technologies. Renewable and Sustainable Energy Reviews. 2021;**141**:110787

[13] Mito MT, Ma X, Albuflasa H, Davies PA. Reverse osmosis (RO) membrane desalination driven by wind and solar photovoltaic (PV) energy: State of the art and challenges for large-scale implementation. Renewable and Sustainable Energy Reviews. 2019;**112**: 669-685

[14] Xu H, Ji X, Wang L, Huang J, Han J, Wang Y. Performance study on a small-scale photovoltaic electrodialysis system for desalination. Renewable Energy. 2020;**154**:1008-1013

[15] Shafaghat AH, Eslami M, Baneshi M. Techno-enviro-economic study of a reverse osmosis desalination system equipped with photovoltaic-thermal collectors. Applied Thermal Engineering. 2023;**218**:119289

[16] Calise F, Cappiello FL, Vanoli R, Vicidomini M. Economic assessment of renewable energy systems integrating photovoltaic panels, seawater desalination and water storage. Applied Energy. 2019;**253**:113575

[17] Ali A, Almutairi K, Padmanaban S, Tirth V, Algarni S, Irshad K, et al. Investigation of MPPT techniques under uniform and non-uniform solar irradiation condition–A retrospection. IEEE Access. 2020;**8**:127368-127392

[18] Yap KY, Sarimuthu CR, Lim JMY. Artificial intelligence based MPPT techniques for solar power system: A review. Journal of Modern Power Systems and Clean Energy. 2020;**8**(6):1043-1059

[19] Yatimi H, Ouberri Y, Aroudam E. Enhancement of power production of an autonomous pv system based on robust mppt technique. Procedia Manufacturing. 2019;**32**:397-404

[20] Saha A, Sultana F, Haque MA, Nasif SS. Design of MPPT mounted solar based double coil DC electric cooker with smart temperature control device. In: 2019 5th International Conference on Advances in Electrical Engineering (ICAEE). IEEE; 2019. pp. 212-217

[21] Mohapatra A, Nayak B, Saiprakash C. Adaptive perturb & observe MPPT for PV system with experimental validation. In: 2019 IEEE International Conference on Sustainable Energy Technologies and Systems (ICSETS). Bhubaneswar, India: IEEE; 2019. pp. 257-261. DOI: 10.1109/ICSETS.2019.8744819

[22] Mohammed R, Moussaoui NE, Kassmi K, Malek R, Deblecker O, Bachiri N. Autonomous solar cooker with photovoltaic energy. Institut Marocain de l'Information Scientifique et Technique IMIST, Revues Scientifiques Marocaines. New Journal for Science Vulgarization. 2021;**1**(2):54-62. ISSN: 2028-7181. Available from: www.supmit.org, https://revues.imist.ma/index.php/NJSV/

[23] Hafeez MA, Naeem A, Akram M, Javed MY, Asghar AB, Wang Y. A novel hybrid MPPT technique based on Harris Hawk optimization (HHO) and perturb and observer (P&O) under partial and complex partial shading conditions. Energies. 2022;**15**(15):5550

[24] Mousa HH, Youssef AR, Mohamed EE. State of the art perturb and observe MPPT algorithms based wind energy conversion systems: A technology review. International Journal of Electrical Power & Energy Systems. 2021;**126**:106598

[25] Yu K et al. A performance-guided JAYA algorithm for parameters identification of photovoltaic cell and module. Applied Energy. 2019;**237**:241-257

[26] Cuce E, Cuce PM, Bali T. An experimental analysis of illumination intensity and temperature dependency of photovoltaic cell parameters. Applied Energy. 2013;**111**:374-382

[27] Al-Subhi A. Parameters estimation of photovoltaic cells using simple and efficient mathematical models. Solar Energy. 2020;**209**:245-257

[28] AbdElminaam DS et al. An efficient heap-based optimizer for parameters identification of modified photovoltaic models. Ain Shams Engineering Journal. 2022;**13**(5):101728

[29] Ben Hmamou D et al. Particle swarm optimization approach to determine all parameters of the photovoltaic cell. Materials Today: Proceedings. 2022;**52**: 7-12

[30] Mrabti T, El Ouariachi M, Tidhaf B, Kassmi K. Caractérisation et modélisation fine du fonctionnement électrique des panneaux photovoltaïques. Revue des Energies Renouvelables. 2009;**12**(3):489-500

[31] Mrabti T, El Ouariachi M, Tidhaf B, Kassmi K, Chadli E, Kassmi K. Modélisation des propriétés électriques et caractérisation des panneaux photovoltaïques. Journal of Renewable Energies. 2009;**12**(1):105-116

Chapter 5

Solar Energy Conversion Efficiency, Growth Mechanism and Design of III–V Nanowire-Based Solar Cells: Review

Fikadu Takele Geldasa

Abstract

Nanowires (NWs) are 1D nanostructures with unique and wonderful optical and electrical properties. Due to their highly anisotropic shape and enormous index of refraction, they behave as optical antennae with improved absorption and emission properties, and thus better photovoltaic cell efficiency compared to a planar material with equivalent volume. Implying important advantages of reduced material usage and cost as well as due to its direct bandgap and its flexibility for designing solar cells, we choose to review III–V NWs. Their bandgap can easily be tunable for growing on the cheapest Si substrate. The recent developments in NW-based photovoltaics with attractive III–V NWs with different growth mechanisms, device fabrication, and performance results are studied. Recently, III–V NW solar cells have achieved an interesting efficiency above 10%. GaAsP NW has achieved 10.2%; InP NW has achieved 13.8%; GaAs NW has achieved 15.3%; and moreover the highest 17.8% efficiency is achieved by InP NW. While the III–V NW solar cells are much more vital and promising, their current efficiencies are still much lower than the theoretically predicted maximum efficiency of 48%. In this review, the chapter focused on the synthesis processes of III–V nanowires, vapor-liquid-solid growing mechanisms, solar light harvesting of III–V nanowire solar cells, and designing high-efficiency and low-cost III–V nanowire solar cells.

Keywords: III–V nanowires, nanowire design, nanowire synthesis, photovoltaic, solar cells

1. Introduction

Energy can be added to the basic need of humans to live on the earth's planet. We need energy in our daily life and economic development, but there is an insufficient energy demand in our world especially for developing countries [1]. The demand for energy is increasing exponentially due to the global population growth and economic development. As the United Nations Department of Economic and Social Affairs (UNDESA) has reported, population size is predicted to extend by two billion within the next 30 years. The expansion rate of the world population indicates that the present world population could jump from currently 7.7 billion to 8.5 by 2030, 9.7 billion by 2050, and 10.9 billion by 2100 [2]. For this population expansion, enormous energy will be required. However, fulfilling this energy demand is a key challenge and a huge obstacle for dreaming of continuous green earth [3]. Currently, fossil fuel-based energy is dominating worldwide, which is meant since it is not replaceable it is running out very fast. In addition to this, to control the amount of CO_2 within the air, it is necessary to reduce the energy demand from fossil fuels and increase the supply of the energy from renewable energy sources [4]. As an alternative to fuel energy, and to minimize CO_2 emission, solar cells, among all the renewable energy resources, can provide an efficient and environmentally friendly solution, for a sustainable green earth, which converts sunlight directly into electricity [5, 6]. The amount of energy humans use annually is about 4.6×10^{20} joules, and this amount of energy is delivered to Earth by the Sun in 1 hour [7]. The largest power that the sun unceasingly delivers to earth is 1.2×10^5 terawatts, which is bigger than each different energy supply, either renewable or nonrenewable [8]. It dramatically exceeds the speed at which human civilization produces and uses energy currently about 13 TW [2, 8, 9]. Depending on the estimation of the population growth rate; the global energy demand is predicted to exceed 30 terawatts by 2050, about double the current energy [2].

Solar cells have been widely utilized in different replaceable energy generation projects including roof-top installations, solar farms, spacecraft, and portable solar battery banks [10]. More importantly, solar cells have been also utilized in building-integrated photovoltaic systems for harvesting solar power, toward the goal of self-sustainable modern infrastructures, such as glass-greenhouses, bus stops, and smart building components, that is, energy generating and saving PV glass [11]. Although the resource potential of photovoltaic (PV) is gigantic, it currently constitutes a little fraction of the worldwide energy supply. One among the factors limiting the wide-spread adoption of PV is its low-energy density, low efficiency, and comparatively high-cost as compared to other energy technologies [12, 13]. In order to widely apply PV, scientists and researchers around the world are still conducting research on this area, including the event of varied sorts of solar cells that specialize in improving the conversion efficiency also [14]. One among the foremost relevant metrics for PV devices is that the power conversion efficiency (PCE), that is, the efficiency with which sunlight is often converted to electric power. There are several factors, from structural defects to resistance to shading effects, which affect the conversion efficiency, also as the overall performance of solar cells.

A significant effort in photovoltaic research today is objectively to enhance PCE, while simultaneously reducing cost [15]. The overwhelming majority of today's PV market consists of three types of generation [16]: the first-generation PV is silicon-based solar cell modules, which currently dominate the solar power market due to their low-cost and long-term reliability, but only convert about 8–19% of the available solar power [17]. Second-generation PVs are thin-film solar cells that aim to decrease cost by utilizing less material and depositing on inexpensive substrates, such as metal foil, glass, and plastic. This type of PVs includes cadmium telluride (CdTe), amorphous Si, and copper indium gallium diselenide (CIGS), all of lower material quality and PCE compared to first-generation cells [18]. In order to overcome these shortage, third-generation PVs [19] are recently being pursued that aim to strike the Shockley-Queisser efficiency limit of ~30% (1 Sun) for one p-n junction [20], while keeping or reducing cost.

The III–V multi-junction planar solar cells are included under third-generation PV and have attracted several interests in the recent candidate of the solar cells that have terribly high efficiencies larger than 40% grown on Ge substrates [21]. Nevertheless, planar III–V materials and Ge substrates needed for these devices are too rare and expensive for widespread use [22, 23]. Since the main barriers to the large-scale uses of solar energy are due to the difficulties in balancing the cost and efficiency of existing devices, innovations are needed to reap solar power with greater efficiency and economic viability. The right resolution is to form the high-efficiency III–V solar cells onto the cheap mature Si platform and develop III–V/Si two-junction cells [22, 24]. It has been foreseen that III–V/Si are able to achieve an efficiency of above 40%, nevertheless, the lattice and thermal expansion coefficient mismatches between III–V layers and Si substrates are still preventing the effective implementation of this idea [22].

By reducing the size of materials from bulk to nanoscale and developing the cheap growth method can solve the problem in III–V multi-junction thin-film solar cells. Recently solar cells in one dimension and zero dimensions geometry materials have got attention. Different materials in nanowire geometries, such as Si, III–V compounds (e.g., GaAs, InP, and III-nitride-based), II-VI compounds (e.g., CdS/CuS_2 and CdS/CdTe), and most recently perovskites have been studied for solar energy harvesting [25]. NW-based solar cells are forest or single of one dimensional (1D) rods, wires, or pillars having lengths typically on the order of microns and diameters on the order of tens to several nanometers. They have unique and wonderful optical and electrical properties and they also offer flexibility to create heterojunctions in both axial and radial directions. Due to their highly anisotropic shape and enormous index of refraction, they behave as optical antennae with improved absorption and emission properties, and thus better photovoltaic cell efficiency compared to a planar material with equivalent volume [26]. The theoretical efficiency of an NW array solar cell can reach approximately 32.5% for bandgap at approximately 1.34 eV under AM 1.5 solar spectrum, exceeding that of a planar bulk solar cell (31%) with the same bandgap, implying an important advantage of reduced material usage and cost [27, 28]. The theoretical power conversion efficiency of 48% is also reported using $Al_{0.54}Ga_{0.46}As$, GaAs, and $In_{0.37}Ga_{0.63.}$ As NWs arrays grown on a silicon substrate [29].

The NW arrays could also provide substantial reductions in material consumption as well as production costs for III–V-based solar cells, in part because they can be grown on low-cost substrates, such as silicon [30]. Among III–V-based solar cells, GaAs and InP are of specific interest for photovoltaic cell applications due to their direct bandgaps, which are close to the ideal value for maximizing PCE under AM 1.5G spectrum [31]. For the first, the best-reported efficiency above 10% for III–V NW is an InP nanowire cell with 13.8% efficiency [30]. The highest power conversion efficiency of the III–V NWs record is caught by InP NWs, which is 17.8% [32]. This value is approached to a planar solar cell that has been reported for silicon radial junction with vertically aligned tapered microwires achieving power conversion efficiency of 18.9% [33]. The principal goal of third-generation PVs is not only the continual increase of power conversion efficiencies but also the reduction of solar cell development costs; novel hybrid materials can provide a practical solution [27]. The array nanowire solar cells are much more important and promising. However, their current efficiencies are much lower than their theoretical prediction. NW synthesis, characterization, and device fabrications are the challenges to achieve the theoretical efficiency predicted theoretically [34].

In this review, we have focused on the synthesis process of III–V nanowires, solar energy harvesting, photon-generated carriers, different design of nanowire solar cells, and ultimately the mostly achieved power conversion efficiency for some of III–V NWs. III–V NW solar cells have gained attention, especially since 2009 and many papers have been published. Depending on these published papers, we have discussed the papers published since 2010 for each aforementioned focused area in this review.

2. Nanowires synthesis

Nanowire can be synthesized through three approaches: i) top-down approaches, ii) bottom-up approaches, and iii) the combination of top-down and bottom-up approaches.

2.1 Top-down approach

The top-down approach begins with a bulk material (microscopic materials), which will be by selection removed to create NWs through lithography patterning and wet/dry etching method [35]. From epitaxially grown-up thin films, they provide the advantage of fabricating NWs with exactly controlled doping profile and layer thickness [23]. If this NW structure has a p-n junction, it will be incorporated as an axial p-n junction after the NWs are formed. To create a radial p-n junction, ion implantation and molecular monolayer doping (MLD) can be used [31, 36, 37]. In the fabrication of nanowires, numerous lithographic styles are used with controllable exposure, size, and distance for dependable light-trapping and latterly high-effectiveness solar cells [38]. The traditional optical lithography can offer a high result, but its essential dimension is confined by the optical phenomenon restriction of the sun wavelength [38, 39]. On the other hand, traditional electron-beam lithography (EBL) has a veritably high resolution but suffers from high-cost and low throughput [40, 41]. Nanoimprint lithography (NIL) [42, 43] can be used in order to obtain both high throughput and resolution, and self-powered parallel electron lithography may be used [44, 45]. The NIL technology avoids light diffraction in optical lithography and can fabricate the nanowires with fabricating accuracy up to several nanometers [46]. The process includes lithography patterning, and then dry etching to obtain nanowires with vertical and smooth sidewalls [36, 47]. Subsequently, wet etching processes are also conducted to first etch the remaining etching masks followed by the removal of physically damaged and nonstoichiometric oxidized surface layers [48]. The top-down have disadvantages when compared with bottom-up approaches. It does not offer any material saving and also lack freedom in material design. Furthermore, the etching process could introduce surface defects that adversely affect the nanowire's optical and electrical properties, and thus lead to much-degraded device performance [49].

2.2 Bottom-up approach

Bottom-up approach NW synthesis is supported by gas-phase epitaxial growth technique to supply detached NW ensembles with or without order. There are many techniques employed under the bottom-up approach for nanowire growth, such as chemical vapor deposition (CVD) [46, 50, 51], chemical-beam epitaxy (CBE) [52, 53], laser ablation [54], and hybrid vapor-phase epitaxy [55, 56]. Nevertheless, III–V semiconductor nanowires are mainly grown by either metal–organic vapor-phase epitaxy (MOVPE) [57–60] or molecular-beam epitaxy (MBE) technique [61, 62] with and without catalysis assistance. Catalyzed growth involves the use of metal nanoparticles, such as Au, Al, and other metals [34].

The catalysts that are used as an assist in the growth of NWs can be external or from the elements of materials used to grow NWs, which are called seed particles. In general, we can classify the NWs growth mechanisms into four as shown in **Figure 1**:

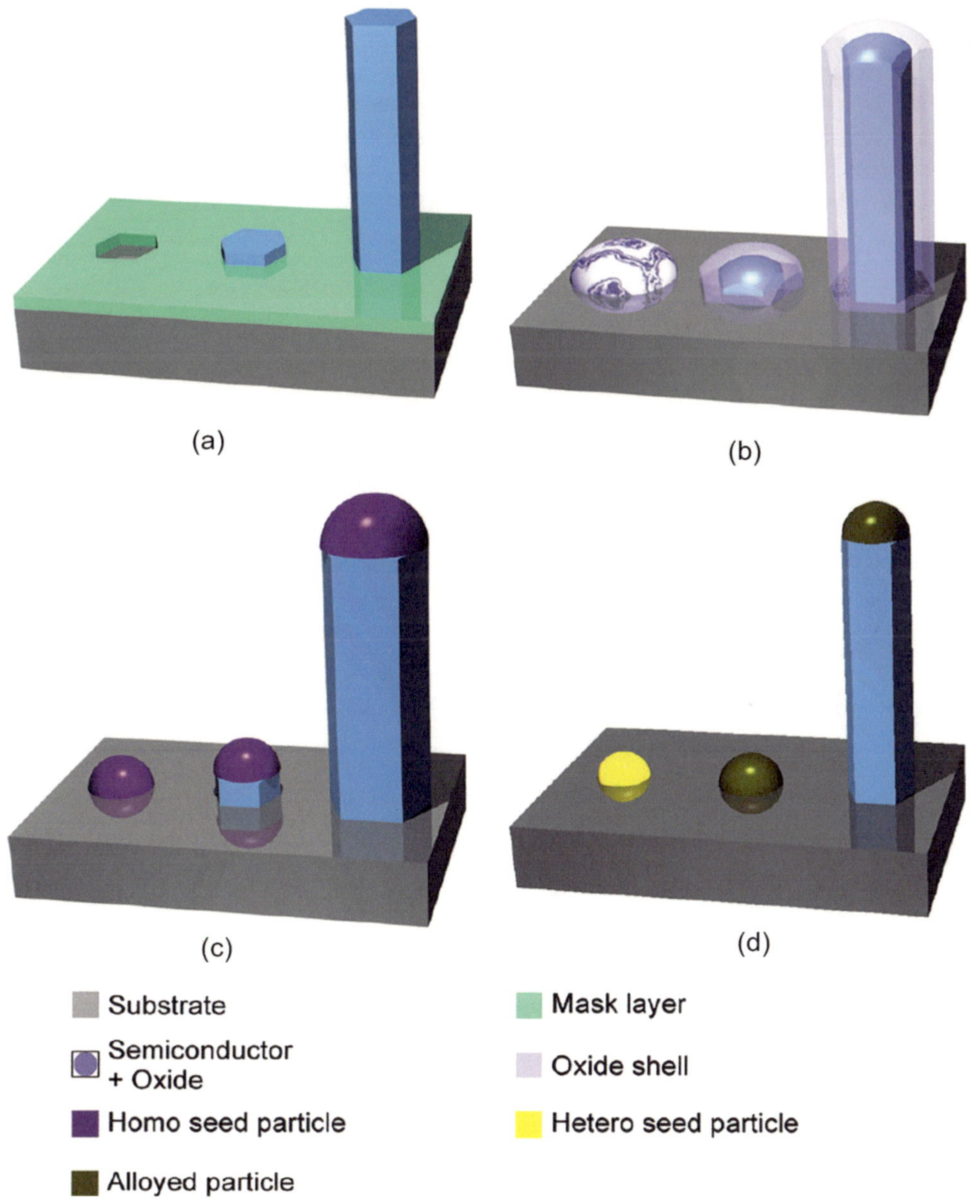

Figure 1.
Schematic representation of four basic nanowire growth mechanisms: (a) selective area epitaxy, (b) oxide-assisted growth, (c) homoparticle growth, and (d) heteroparticle growth [63].

(i) homoparticle growth, (ii) heteroparticle growth, (iii) non-catalyst growth, and (iv) oxide-assisted growth. The seed particles can be homoparticle growth (**Figure 1c**); in this case, a seed particle is formed consisting of one or all elements used for wire growth or it can be simply a self-assisted growth. As the seed particle size varies during growth both length and diameter increase. The seed particles can also be heteroparticle growth (**Figure 1d**) and in this case, a seed particle (typically Au) is deposited prior to growth, in simple words, it is a foreign metal-assisted growth technique. Throughout heating to increase temperature the seed particle alloys with the substrate and/or material forms the gas phase. In this case, particle size during growth is constant. Noncatalyzed growth includes selective area epitaxy (SAE) where growth occurs on a prepatterned substrate [64–66]. In selective area epitaxy, an epitaxial layer nucleates in openings of a mask layer and continuously grows in height; its lateral growth is restricted by low-energy facets (**Figure 1a**). [63, 67]. Oxide-assisted growth (OAG) is additionally a mechanism used in crystal growth with the aid of the semiconductor substance's oxides as a passivating shell to suppress the subsequent growth [68]. During OAG growth, the semiconductor and

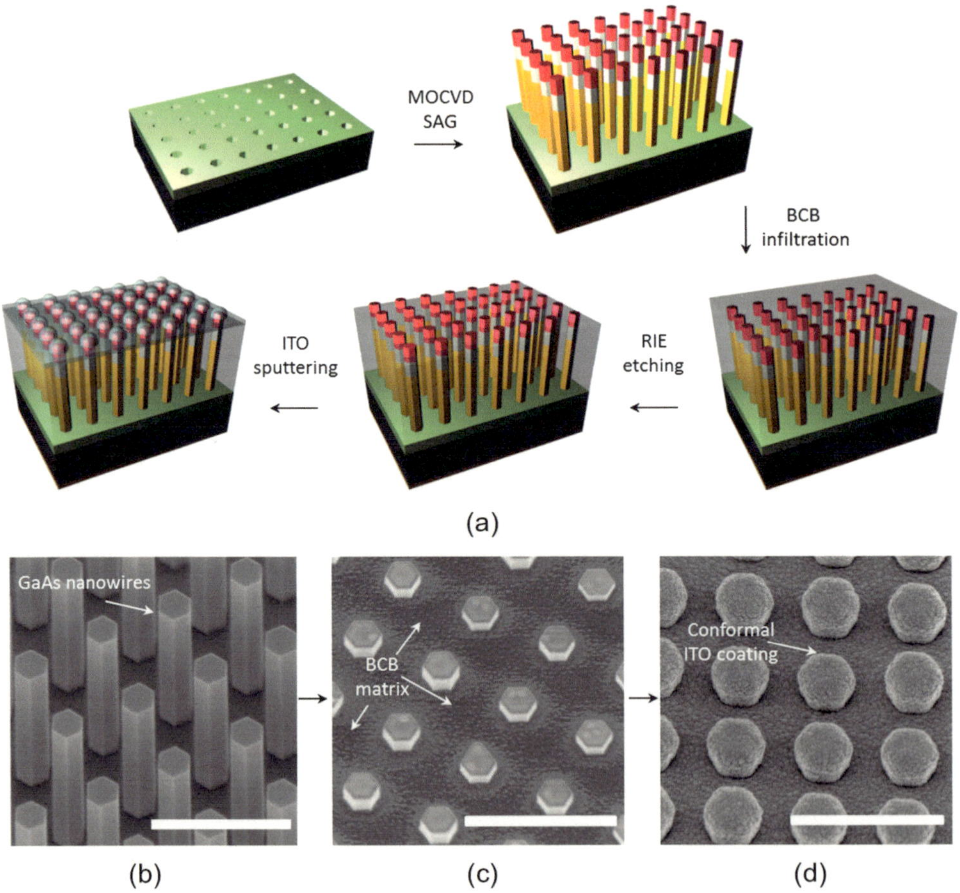

Figure 2.
Solar cell fabrication process and SEM images. (a) Fabrication steps of GaAs nanowire array solar cells with axial junction, (b) 30° tilted SEM image of as grown vertical GaAs nanowire array on GaAs (111) B substrate, (c) SEM image after nanowires are embedded in BCB and etched by RIE to expose short tips, and (d) SEM image after coating of ITO film by sputtering. A conformal dome-like cap is formed on the tips of nanowires [69].

its oxide are adsorbed on the substrate, where the semiconductor produces nucleation centers, which also create the semiconductor nanowires, while the oxide forms a passivating shell (**Figure 1b**).

For the non-catalyzed growth method, Maoqing Yao et al. [69] fabricated arrays of GaAs nanowires solar cell with an axial p-i-n junction, which are grown by selective area growth (SAG) method, using mass production compatible metal-organic chemical vapor deposition (MOCVD) technique. This growth method is free of the metal catalyst. The fabrication process of their axial junction GaAs nanowire solar cell is shown in **Figure 2a**. Their fabrication steps are (1) electron-beam lithography is applied to form hole array in silicon nitride mask, (2) SAG of p-i-n GaAs nanowire using MOCVD, (3) BCB infiltration, (4) reactive ion etching (RIE) to expose nanowire tips, and (5) transparent conductive indium tin oxide (ITO) deposition. The SEM images for as-grown vertical GaAs nanowire array (**Figure 2b**), after nanowires are embedded in BCB and etched by RIE to expose short tips (**Figure 2c**) and after coating of ITO film by sputtering (**Figure 2d**) is displayed.

3. Vapor-liquid-solid growth mechanism

NWs are grown from the vapor-liquid-solid (VLS) method during which an NW is grown from the vapor phase employing a metal seed particle, like Au as shown in **Figure 3a**, that is usually liquid at the expansion temperature (after alloying with the substrate and/or growth species) [56, 70, 71]. The VLS growth technique as its name suggests is the growth method from the combination of the three phases (vapor phase, liquid phase, and solid phase). The vapor phase is the gas phase precursor, the liquid phase is the catalyst and the solid phase is the final grown nanowire. A foreign metal can be used to promote NW growth by forming a liquid eutectic with the desired NW material through this mechanism (**Figure 3a** and **b**). In such NW synthesis, the chemical precursor's vapors are transported into the hot zone by an inert carrier gas and react on a substrate with metal catalyst nanoparticles [72]. With the proper choice of substrates, catalysts, precursors, and growth conditions, various types of vertical NWs, as well as planar NWs can be achieved [73]. While VLS synthesis is the most common with the seed particle in a liquid state, a mechanism is also possible if the catalysts remain solid and do not form a eutectic with the NW

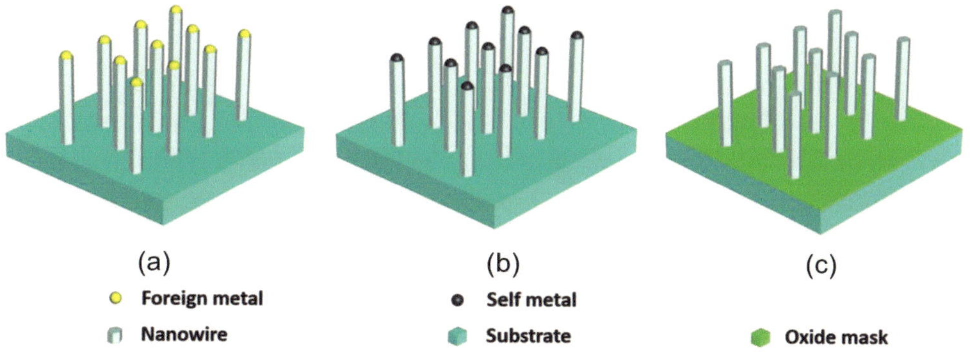

Figure 3.
Schematics for three typical nanowire growth mechanisms: (a) foreign metal–catalyzed growth, (b) self-catalyzed growth, and (c) selective area epitaxy [31].

material [74]. Such solid-phase diffusion mechanism happens below the catalyst's eutectic point with the metal seeds remaining as solid [75]. Ingvar Aberg et al. [76] have grown GaAs NW array solar cells by the VLS growth Au-assisted method, which demonstrated a 1-sun independently verified solar energy conversion efficiency of 15.3%.

As explained in the previous section the VLS method can also be use self-assisted. The self-assisted method uses a component of the NW itself as shown in **Figure 3b**, which avoids the utilization of foreign metal elements [41]. A foreign particle which found on the top of NWs may cause harmful effects like contamination of the NWs, increased contact resistance, or reflection of sunshine from a photovoltaic device [77]. These foreign metallic particles can be etched after device processing, but etching might cause problem and will have a negative effect on the performance of device. **Figure 3c** shows the oxide-supported growth mechanisms.

Mandl et al. [63] have grown InAs NW by a VLS mechanism employing a liquid In droplet and they identified that the presence of the oxide layer is vital to immobilize In

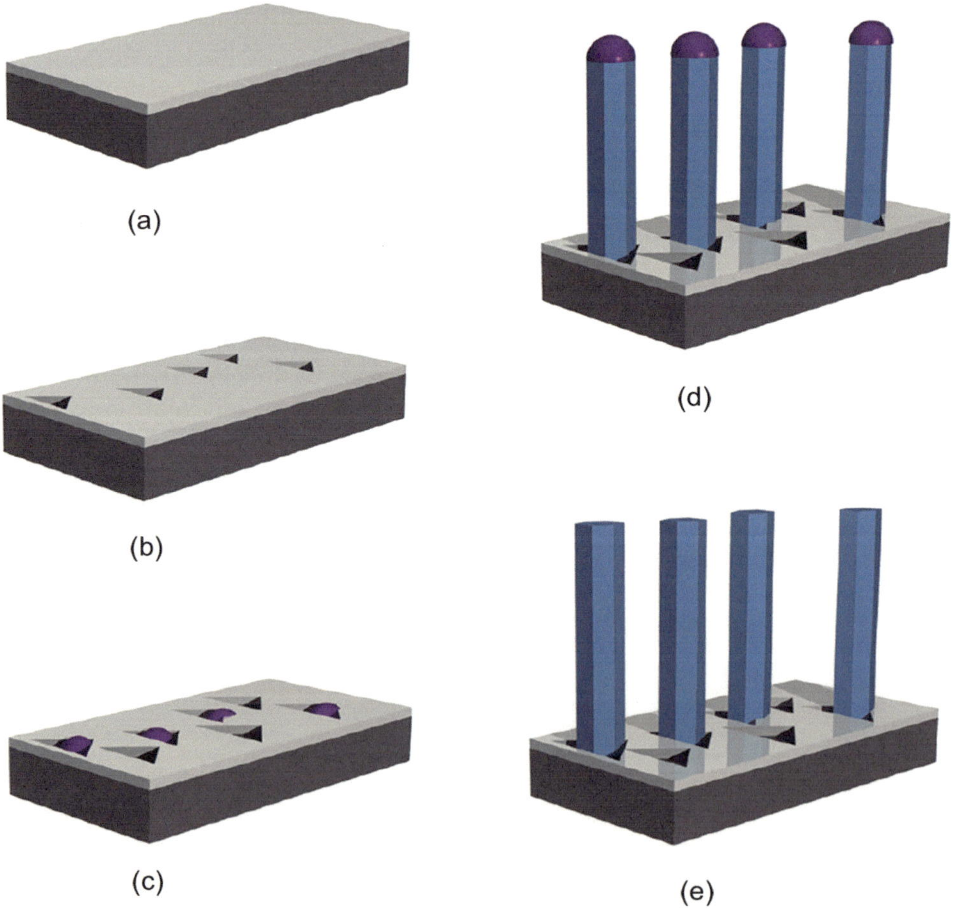

Figure 4.
The growth mechanism illustration of nanowires: (a) layer before growth, (b) layer during heating and creation of hole, (c) formation of In droplets, (d) growth of the NWs under the In droplets, and (e) droplet solidification during cooling [63].

droplets on the surface, restricting the particle size and NW nucleation formed. They have concluded that NWs can be grown in the sequence, as illustrated in **Figure 4**: (a) in the deposited SiO_x openings form during heating the substrate; (b) when the trimethylindium (TMI) precursor is activated, the In atoms adsorbed on the surface diffuse into immobile In droplets formed in the openings of the SiO_x layer; (c) at the interface between indium particles and the substrate underneath, NW growth is began by enhancing the rate of growth in one direction; and (d) after deactivation of the TMI supply, the droplet forms InAs NW.

4. Vapor-liquid-solid growth mechanism

The main challenge regarding the performance of thin-film photovoltaic cell structure is the sunshine reflection losses, for instance, with no treatment materials, around 30% of the light illuminated at the Si surface can be lost due to the reflection at the interface between air and Si [78]. During the illumination of sunlight to the surface thin-film cell, some of the light is converted into energy, others will be transmitted, whereas some parts reflect [79]. The loss due to reflectance can be reduced by using techniques, such as coating with anti-reflection and light trapping materials [80]. To reduce this loss, the most commonly used is dielectric antireflection coatings; however, it is difficult to hide the whole absorption wavelength range. Broadband antireflection methods will be achieved by light trapping schemes, such as inverted pyramid structures, but these boost the cost due to their complicated fabrication method. Contrary, NW arrays have a strong antireflection ability with superior wavelength, polarization, and angle-dependent properties compared to planar structures because NWs can form graded-refractive index layers [80]. Consequently, it will reduce the light reflectance at the interface of the two media by avoiding abrupt changes in the refractive index [81].

Wu et al. [82] have presented a model for effective and fast design of both squarely and hexagonal InP NW arrays to achieve the highest light-harvesting for PV application, achieving the maximal short-circuit current density of 33.13 mA/cm^2. They have investigated the geometrical dimensions for vertically aligned single, double, and multiple diameters of NW arrays. NWs and nanorods have almost the same properties and solar cells can also grow as nanorods morphology to harvest highly efficient sunlight by reducing reflection. Diedenhofen et al. [50] grew layers of GaP nanorods on AlInP/GaAs substrates. They found that nanorods can greatly reduce the reflection and increase the sunshine transmission into the substrate overbroad spectral and angular ranges due to the graded index of refraction. Strudley et al. [58] studied the sunshine transport inside an NW mat. They found that due to mesoscopic transport the high-density semiconductor NW mats exhibit huge interference contributions. From their statistical analysis of intensity oscillations, they linked that transport for focused illumination is governed by a minimum of around three open transmission modes, which is a record low value for light in a 3D medium.

Additionally, semiconductor NW is a 1D nanostructure, which is usually on the order of the sunshine wavelength. Due to their high refractive index, they behave as optical antennae that can modify the absorption and emission properties [80]. The absorption properties of NWs when they are vertically standing are determined by the waveguide modes [82]. InP NW arrays, which are vertically aligned and grown on a semi-infinite SiO_2 substrate are schematically shown in **Figure 5** with either squarely or hexagonal arrangement. Repeatable unit cells in **Figure 6a** and **b** insets

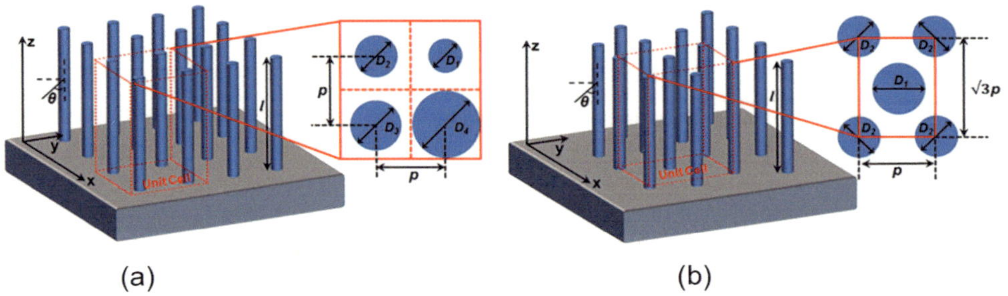

Figure 5.
Schematics of vertically aligned InP NW arrays. (a) Squarely, and (b) hexagonal NW arrays with insets explaining their respective unit cells [82].

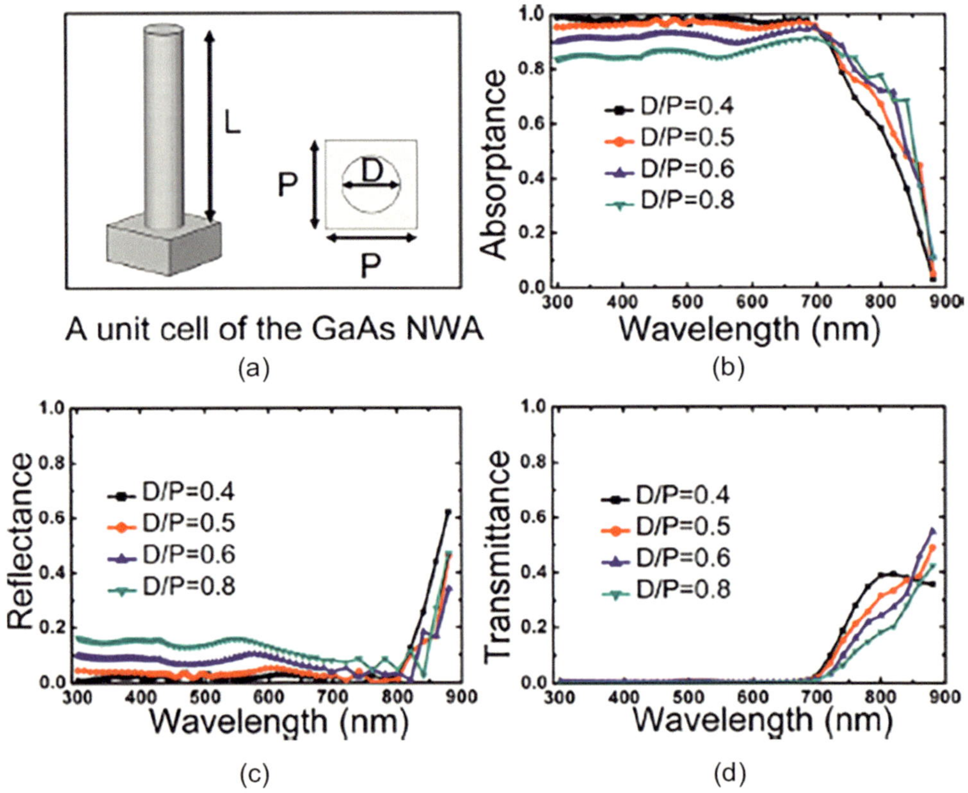

Figure 6.
(a) Schematic drawing of the periodic GaAs NWs structure, (b) absorptance, (c) reflectance, and (d) transmittance of GaAs NW array with different fill factors [83].

show respective characterization dimensions for each arrangement. Such morphology and topology of the NW arrays are in accord with the majority of the InP NW-based photovoltaic cell structures. Within each of the unit cells, the NWs own identical or different diameters as D_i (where i = 1, 2, 3,...). Periodicity P is the core to core spacing of a pair of adjacent NWs that has an analogous value for squarely arranged NWs, whereas fully different values for hexagonal NW clusters.

NWs are more efficient in light absorption compared to thin-film materials of an equivalent volume. Krogstrup et al. [84] have observed a remarkable increase in absorption in single-NW solar cells, which is related to the vertical configuration of the NWs and to a resonant increase in the absorption cross-section, and the results obtained opened a new route to third-generation PVs cells. Their short-circuit current result of 180 $mAcm^{-2}$ is higher than that predicted by the Lambert-Beer law.

On other hand, when the NW is lying horizontally, the absorption properties are determined by leaky-mode resonances, which provide a chance to engineer the light absorption in NWs by controlling their physical dimensions [85]. Once the resonant modes are supported by the NWs leaky, the overlap between the incident electromagnetic attraction field and the guided mode profile is maximized, facilitating enough coupling with incident light.

Due to their outstanding advantages, NW arrays have advanced light trapping ability and hence strongly enhanced optical absorption in comparison with the thin-film [86]. This can significantly enhance the broadband light absorption over a good range of incident angles, especially the near and below bandgap absorption [81, 87]. With the same thickness as thin-film layers, the NWs short-circuit current can reach high results [88].

4.1 Effect of NW diameter and period on absorption

Nanowire diameter and separation are typically on the order of the wavelength of sunlight, where interference effects are dominant, therefore, the reflectance, absorptance, and transmittance of nanowire arrays must be determined using wave optics [89]. Long Wen et al. [83] have simulated to evaluate the efficiency limits of GaAs NW array solar cells and determined the requirements of the optical design for improving the efficiency **Figures 6a–d**. They have suggested that the optimized design NW might absorb 90% of above bandgap sunlight. Their combined optoelectronic simulation results reveal that optimization of optical geometry can lead to an attainable photovoltaic efficiency of 22%.

By fixing the filling factor, which is given by D/P, where D is the diameter of the nanowire and P is the separation between the grown nanowires, the effect of NWs diameter can be determined by varying it. **Figure 7a** shows the optical characteristics of the GaAs NW array with different diameters at a fixed D/P of 0.5 is plotted. The absorptance of a 2.2 μm GaAs thin-film is also plotted for comparison. In short wavelengths, it can be observed that the absorptance spectra for all NW arrays are kept above 90%, which is much higher than the planar case due to the lower effective refraction index, and thus lower reflection at the top of NW arrays. In the long-wavelength regime, the absorption spectra show a significant increase when increasing diameter from 60 to 180 nm. The electromagnetic field can be coupled efficiently into the NWs at resonances, due to the large refractive index contrast between the NWs and surrounding air. For small diameter GaAs NW arrays, with fewer supporting modes, most of the incident light cannot be guided into the NWs. The absorbance of the NW array with D = 180 nm is high above the band gap wavelength, whereas when D is increased to 240 nm the absorbance of the NW array decreased (**Figure 7a**) due to the increased reflection and the insufficient field concentration at longer wavelength.

Figure 7b shows plots of the vertical cross-section of the photogeneration profiles. The NWs with D = 60, 180 nm and D/P = 0.5 under 1$mWcm^{-2}$ sunshines at different

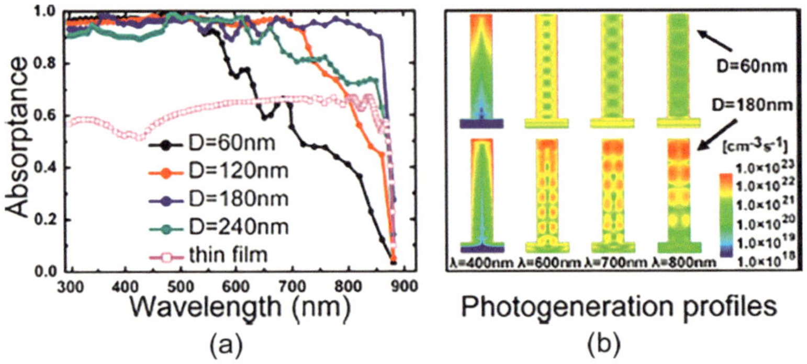

Figure 7.
(a) Absorptance of NWA with different diameters, and (b) Photogeneration profiles calculated by FDTD simulations [83].

wavelengths for photogeneration rates are revealed. At λ = 400 nm, it is concentrated near the highest sides of the NW for both diameters. Only a little fraction of the incident wave is transmitted onto the substrate, this will be explained by the short absorption length of GaAs at this wavelength. At 600 nm and above, for a NW, the photogeneration rates are focused on several lobes that form along the NWs for a NW array with 180 nm diameter, indicating strong guided modes confined within the NWs. In contrast, for the case of D = 60 nm, the optical generation becomes more homogeneously covered by the NWs with a longer wavelength. Clearly, the 180 nm diameter NW array induces a much larger optical concentration than the 60 nm diameter one. From both **Figures 6** and 7 one can easily understand the effect of NW diameter on photon energy harvesting.

4.2 Effect of NW length on absorption

Photocurrent density is often further bettered by adding nanowires length (L). **Figure 8** shows the reckoned donation to all photocurrent from the NWs and the substrate during a GaAs NW PV device for an NW periphery of 180 nm and a period of 350 nm (90). Due to the proliferation of NW length, the donation from the nanowire to all photocurrent rises, while the GaAs substrate donation similarly decreases. The uttermost photocurrent of 27.3 $mAcm^{-2}$ is obtained at 5 μm length, is on the brink of the perfect photocurrent density of 29.9 $mAcm^{-2}$ (calculated by integrating the AM1.5G spectrum above the GaAs bandgap). At the optimum NW diameter, spacing, and length of the harvesting properties of III–V NWs can be improved.

In 2015 Nicklas Anttu [28] compared the effectiveness of InP NW assemblage solar cells with the classical InP bulk solar cells. They accounted an NW assemblage of 400 nm periods, 4 μm length, and 170 nm periphery, which may produce 96 of the short-circuit current accessible within the impeccably taking up InP bulk cell. Also, the NW solar cells cast smaller photons than the bulk cell at the identical occasion, which allows for a more open-circuit voltage. They consequently found that NWs longer than 4 μm can really show, despite producing a lower short-circuit current, an efficiency limit of up to 32.5% that is above the bulk cells.

They have predicted the unborn capabilities in affecting both the emission and absorption characteristics of the NW assemblages, for instance, by (1) varying NWs shape, (2) varying the period of NWs, (3) sheeting the NWs with a nonabsorbing

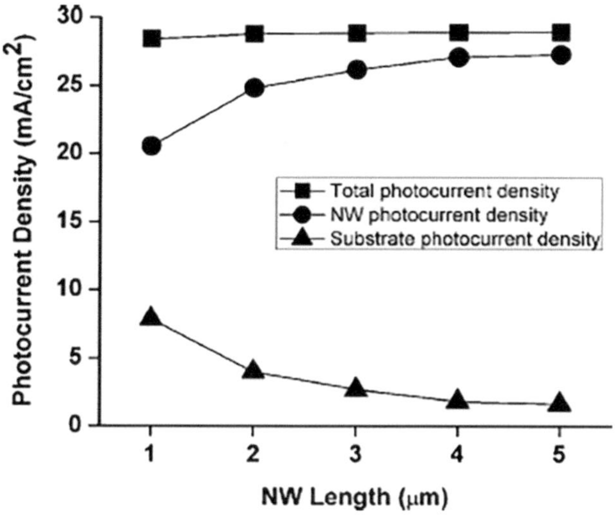

Figure 8.
Theoretical contributions from a GaAs nanowire array and the GaAs substrate to the total photocurrent density in a PV device versus nanowire length obtained at a nanowire diameter of 180 nm and period of 350 nm [90].

dielectric shell, (4) fitting a dielectric material between the NWs, and (5) by introducing optical antireflection layers on top of the NW. Such improvement of the NW array could conceivably further accelerate its effectiveness limit.

5. Absorption depth and diffusion length for III–V NWs

Based on the axis of charge carrier separation, an axial and a radial junction device are the two broadly classifying NW solar cells. The charge carrier separation happens along the length of the nanowire and the radial axis, in axial junction, and radial junction solar cells, respectively. **Figure 9a** and **b** display sunlight absorption and charge carrier separation in both axial and radial junctions NW solar cells correspondingly. In a solar cell, the minimum length needed to attain ample absorption is characterized by absorption depth. The absorption depth explains how deeply light penetrates the NW semiconductor or every type of solar cell device before being absorbed. At the same time, diffusion length describes the maximum length that the minority charge carrier can travel before making recombination non-radiatively [91]. For solar cells, in order for them to efficiently operate, the diffusion length should be higher than the absorption depth, as schematically shown in **Figure 9**. Radial junction is preferable for the fabrication of large-efficiency devices by connecting the light absorption and charge carrier separation axes. In a radial junction PV cell, sunlight absorption is along the main axis of the NW, while the charge carrier separation takes place within the radial direction, which is in nm-scale thickness. In other words, to realize the optimum performance of the NW photovoltaic cell in a radial junction photovoltaic cell, both charge carrier separation, and light absorption can separately be optimized.

Yao et al. [69] have carried out an optical simulation to predict the optimized axial junction and radial NW array for maximum light absorption and have compared the merits and demerits of these NWs. They have also synthesized GaAs NWs

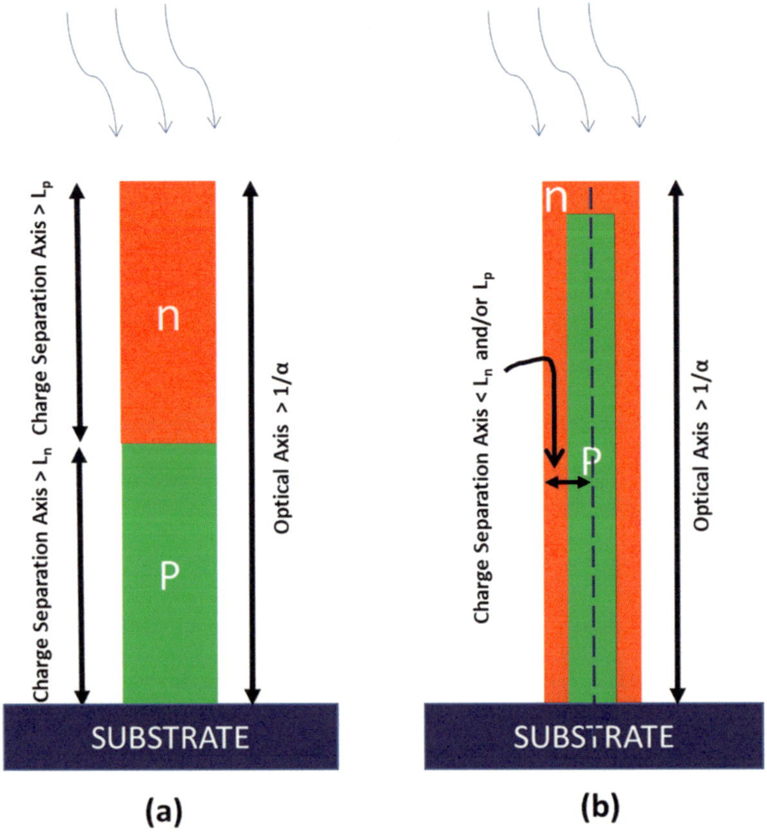

Figure 9.
Nanowires solar cells schematic representation of (a) an axial p-n junction, and (b) a radial p-n junction. α denotes the absorption coefficient of the active material and L_n and L_p denote the electrons and holes diffusion lengths, respectively [91].

solar cells with an axial p-i-n junction by selective area growth method, which is compatible with MOCVD technique, and they observed that low filling ratio NWs are highly absorbed. They have also studied the effect of the diameter and revealed that thicker NWs are favorable because of the high surface recombination velocity on the bare GaAs NW surface. They identified that by decreasing junction depth to around 100 nm and maintaining diameter at 320 nm, able to achieve efficiencies as high as 7.58%. Their results demonstrated that GaAs NWs are good candidates for high-efficiency and low-cost solar energy conversion and open up great opportunities for the next generation photovoltaic based on multi-junction devices composed of lattice-mismatched material systems.

6. Photogenerated carriers of III–V NWs

For solar cells, one of the key needs is to realize efficiency that keeping a huge optical thickness to facilitate high light absorption and a tiny low electrical thickness to facilitate high photogenerated carrier assortment at the contacts. The gathering of high photogenerated carriers depends powerfully on the diffusion length of minority carriers, which decline quickly with the rise in density of defect [22]. Generated carriers are going to be wasted when they are quite one diffusion length far away from the space charge region [92]. The diffusion length, L_d, of electrons or holes in a semiconductor is defined by the mean distance the relevant charge moves within the semiconductor. It is influenced by the mean distance the relevant charge moves within the semiconductor and recombination/extraction from the semiconductor. Diffusion is the movement of charge carriers directed by a concentration gradient. The diffusion coefficient (D) and additionally the equivalent term among the presence of a field, mobility (μ), are associated with one another by the relation [92]:

$$D = \frac{\mu kT}{q} \tag{1}$$

and

$$L_d = (D\tau)^{1/2} \tag{2}$$

where τ is the charge lifetime.

When the cell is not operational at open-circuit voltage, that is, the charge is extracted, and then the lifespan can clearly be less due to the removal of the charge extracted. This is no longer an intrinsic property of the absorbing semiconductor itself, however, depends on the interfaces that exist between the semiconductor and charge extraction phases. The lifespan of charge refers to the minority charge carrier lifespan for semiconductors that are obviously either n-type or p-type. Differentiation into majority and minority carrier lifetimes is not obvious for an intrinsic semiconductor, such as the intrinsic semiconductor in a p-i-n cell [93].

In a conventional thin-film device, the gathering path of the generated carriers is parallel to the solar photon traveling path. Thus, thick enough absorption materials are in high demand on the quality of the crystal, in order that the carriers can easily undergo without any substantial recombination. The morphological anisotropy of nanowires provides the advantage of decoupling the optical and electrical thickness of PV cells by using the co-axial contact structure [91]. It can absorb sunlight along the entire nanowire, while the generated carriers are frequently separated within the radial direction. The radial distance that carriers need to travel (in the 100 s nm range) is generally much lower than, or similar to the minority carrier diffusion length. So far, the orthogonally severed sunshine and carrier separation paths can cause low bulk recombination, and hence high effectiveness. Also, the NWs have a high surface-to-volume ratio, which offers a large junction area that will further enhance the charge separation effectiveness.

The study showed that the influence of adjusting the diffusion length under radial junction may be a smaller amount than in planar junction, that is the utmost efficiency of both radial p-n junction geometry and planar geometry can increase with increasing diffusion length, but the planar geometry increases more [94, 95]. The difference in the performance between the planar and radial structures for III–V semiconductors with a high carrier diffusion length is, not as clear as that for Si [94]. However, NWs have a large surface-to-volume ratio and hence, a large density of surface state [93, 96]. All these merits allow using lower-purity, less expensive materials with low minority carrier diffusion lengths to make high-efficiency solar cells. Consequently, the use of the NW structure can enormously decrease the device cost. Due to these advantages, NWs are promising high-efficiency and less expensive solar cells and have the potential to revolutionize solar power harvesting technology.

7. III–V NWs design for high-efficiency and low-cost solar cells

The distinctive structure and advanced properties of NWs provide additional freedom in constructing novel solar cells with high-efficiency and low-cost. That is solar cells can be designed in different architectural such as tandem solar cells, axial tandem solar cells, multi-terminal solar cells, inorganic nanowire/organic hybrid solar cells, branched solar cells, and flexible solar cells, in which III–V nanowires can also be designed.

Tandem solar cell [97] is one type of design in order to have high efficiencies in solar cells, which is to use multiple semiconductors epitaxially grown on top of each other. **Figure 10** shows the system with two different semiconductor materials, where one material is used as top materials and different materials are used as bottom cell materials. In this figure L_{top}, L_{bot}, D_{top}, and D_{bot} illustrate the length of the top cell, length of the bottom cell, the diameter of the top cell, and diameter of the bottom cell, respectively. It is to absorb high-energy light in a large bandgap top cell in such a tandem solar cell. Compared to the single junction cell, the thermalization loss of the high-energy light is decreased in the top cell. Then, the lower energy light continues to the bottom cell where these energies are absorbed. The bottom cell has a lower bandgap and due to the lower bandgap than in the single-junction cell, more photons are absorbed in the bottom cell. Consequently, the tandem solar cell can absorb more photons than single-junction cells and also can have reduced thermalization loss. However, in planar cells, the crystal lattice constant should be matched in adjacent subcells to offer high-quality materials [97]. Due to this lattice mismatch, they cannot grow on Si substrate, which is the second most abundant earth element and cheap. Moreover, the III–V multi-junction cells in the conventional thin-film structure can give high-efficiency but need to use Ge as substrates, which is expensive. The blending of III–V solar cells on Si substrates can greatly reduce the value, which is extremely challenging.

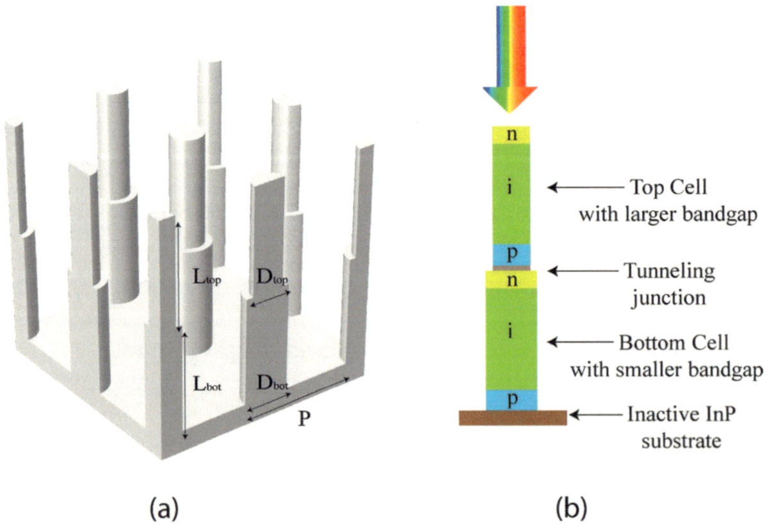

Figure 10.
(a) Schematic representation of a dual-junction NW array on the inactive substrate, and (b) illustration of the electrical design of NWs with axially configured p-i-n junction in which a tunnel junction connects the bottom and the top subcell [97].

Nanowire structures give an obvious advantage for multi-junction solar cells compared with thin-film cells. NWs have efficient strain relaxation, which permits for the fabrication and combination of dislocation-free and highly lattice-mismatched materials. In another word, III–V nanowire arrays can be grown on top of a Si substrate, giving the prospect of using the Si substrate as the bottom cell. **Figure 11a** shows the growing of III–V NWs on Si substrates consisting of a bottom Si cell and a top III–V nanowire cell [99].

The optimum structure needs the absolute stylish NW cell to have a direct bandgap of near 1.7 eV, which can be achieved by employing a number of III–V emulsion semiconductor material systems. The optimum structure also requires equal current from each sub-cell, videlicet a current-corresponding condition. This may be realized by conforming the periphery, length, and period of the NW array. Thus, NW solar cells have further degrees of freedom compared with thin-film solar cells, whose current-matching is achieved by conforming to the consistency of the absorbing subcaste in each subcell. The optimum building needs the absolute stylish NW cell to possess a direct bandgap of closer to 1.7 eV, which can be attained by engaging a number of III–V semiconductor materials. The optimum building also requires equal current from each subcell, namely a current-matching condition. This may be realized by conforming to the length, period, and diameter of the NW array. Thus, NW solar cells have more degrees of freedom compared with thin-film solar cells, whose current-matching is achieved by adjusting the thickness of the absorbing layer. Hu et al. [98] designed the current matching 1.7 eV III–V NW top and 1.1 eV Si planar bottom cell by tuning the NW diameter and period (**Figure 11b**). They obtained the best photocurrent density of 17.8 $mAcm^{-2}$ at NW diameter of 180 nm, period of 350 nm, and length of 5 μm, which result in 89.4% absorption of the AM1.5G spectrum and

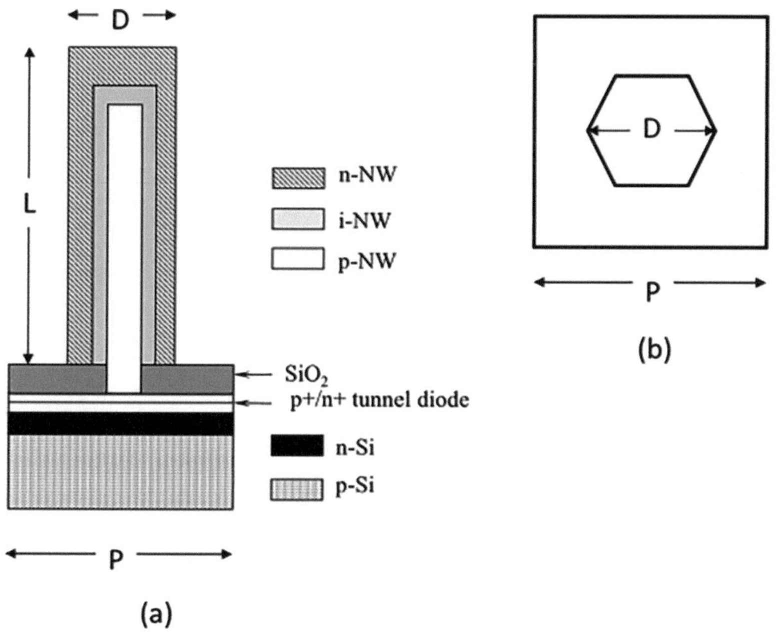

Figure 11.
Model geometry of III–V NW on a Si substrate (a) side view showing doped layers, and (b) top view showing a hexagonal NW of diameter D arranged in a square array of period P [98].

a promising efficiency above 30% under one sun illumination. Yao et al. [100] have reported the growth of III–V NW on Si tandem cells with the GaAs nanowire top cell and the Si bottom cell with a circuit voltage V_{oc} of 0.956 V and a high-efficiency of 11.4%. Their simulation showed that the current-matching condition plays a crucial role in the overall efficiency of the device. They also have characterized that GaAs NW arrays were grown on lattice-mismatched Si substrates, which are less expensive. They concluded that tandem solar cells supported top GaAs nanowire array solar cells grown on bottom planar Si solar cells, open up great opportunities for high-efficiency and low-cost multi-junction solar cells.

Axial and radial tandem solar cells [101–103] are another form of solar cell designing. In axial tandem solar cells, because the photogeneration events happen most often in the middle of NWs, they cannot intrinsically block the generated carriers from reaching the surface and recombining like the radial junctions. Due to a similar reason, the radial tandem solar cell faces a challenge of inefficient absorption for the cell junctions away from the core of NWs. Thus, a composite structure that combines the advantages of the axial and radial structures would provide much higher efficiency compared with homogeneous ones [104].

Furthermore, NWs have a little cross-section, which allows them to accommodate big strains axially and laterally and this may greatly facilitate the blending of materials with large lattice mismatch, providing more freedom within the structure design compared with thin-film devices [105].

An axial NW heterojunction structure with lattice mismatch can be created from the results that the axial junction will distribute the strain across the interface, which will relax the straining step by step and elastically. Regardless of the length, there exists a critical diameter below which no interface dislocation is often introduced. Dislocation-free NWs heterojunctions, such as GaAs/GaP [106], InAs/InSb [107], and InAs/InP [108] have been realized even with large lattice-mismatch. For example, Ercolani et al. [107] have reported the Au-assisted CBE growth of defect-free zinc-blende structure InSb NWs. InSb NW was grown on the upper sections of InAs/InSb heterostructures on the InAs (111) B substrates. They have also observed that zinc-blende structure InSb is often grown without any crystal defects.

With the same concept, the nanowire core has advantages with regard to lattice-mismatch strain in that it can share the nearest mismatch strain, which results in a drastically reduced strain within the shell [109]. NW core-shell structure can thus accommodate larger lattice mismatch compared with thin-film structures [110]. V. Nazarenko et al. [111] have reported the growth of core-shell InGaAs/GaAs nanopillars by MOCVD on Si substrates. They demonstrated that a shell thickness around 160 nm defect-free GaAs grown on $In_{0.2}Ga_{0.8}As$ core NWs despite a large lattice mismatch amounts to 2% for the 20%. Their TEM characterization showed an outstanding crystal quality in the entire pillar without defects. Wang et al. [112] have grown a novel NW structure for solar cells that axially connects core-shell p-n junctions (**Figure 12a**) with different bandgaps. In order to evaluate the performance of this NW, they have used a coupled 3D optoelectronic simulation and their simulation results revealed a high conversion efficiency of 16.8% at a low filling ratio of 0.196. After an outstanding current matching, a promising efficiency of 19.9% was achieved at a low filling ratio of 0.283, which is much higher than the tandem axial p-n junction under the same conditions. **Figure 12b** illustrates vertically aligned NW arrays of axially connected core-shell structures.

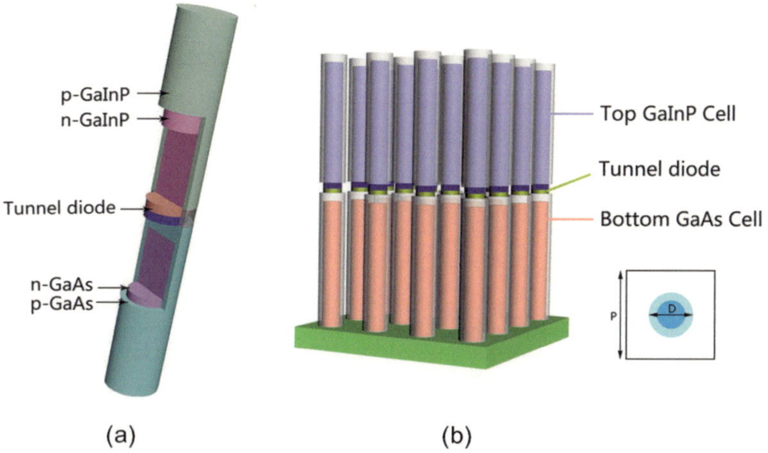

Figure 12.
(a) 3D illustration of axially connected core-shell p-n structure with different III–V materials is axially connected by the tunnel diode in a NW, and (b) Schematic drawing of vertically aligned NW arrays [112].

The unique structure of NW p-n junctions enables substantial light absorption along the NW length and efficient carrier separation and collection within the radial direction. Heurlin et al. [113] demonstrated the growth of tandem junction InP NWs on a Si substrate. By applying in situ etching for total control over axial and radial growth they connected two photocurrents having p-n junctions in series by a tunnel junction. They observed a rise up of V_{oc} by 67%. They also believed that this provides the best way toward realizing high-efficiency multi-junction solar cells that can be fabricated on a large area and low-cost Si substrates.

Multi-terminal NW solar cell is also another promising design of nanowires. Introducing multiple bandgap concepts into NW solar cell designs has high promise for maximum solar conversion efficiency [114]. Dorodnyy et al. [29] have proposed a multi-terminal NWs solar cell design as shown in **Figure 13**. Their NW design resulted in theoretical power conversion efficiency of 48% utilizing an efficient lateral spectrum splitting between three different III–V material NW arrays grown on a flat silicon substrate. These authors used $Al_{0.54}Ga_{0.46}As$, GaAs, and $In_{0.37}Ga_{0.63}As$ NWs with bandgap 2.01, 1.42, and 0.93 eV, respectively. However, the main challenge would be the matter of growing different NW groups with different lengths required for device fabrication.

The mixing of inorganic NW and organic will give the opportunity to have hybrid solar cells and is also another design of the solar cells to offer high-efficiency materials [115–118]. These two materials have their own advantages. Inorganic materials commonly possess high carrier mobility and affinity, whereas organic polymers commonly possess low carrier mobility and a short lifetime, which leads to low device efficiency. However, organic polymers are low in cost; as a result, researchers attempt to mix together the advantages of the two material systems. Due to the fast and efficient charge separation or collection, a greatly enhanced efficiency is hence expected for the inorganic NW/polymer combination. H Bi and R R LaPierre have fabricated hybrid solar cells consisting of GaAs NW arrays and poly(3-hexylthiophene) or P3HT. They have been fabricated by spin-coating poly(3-hexylthiophene) (P3HT)

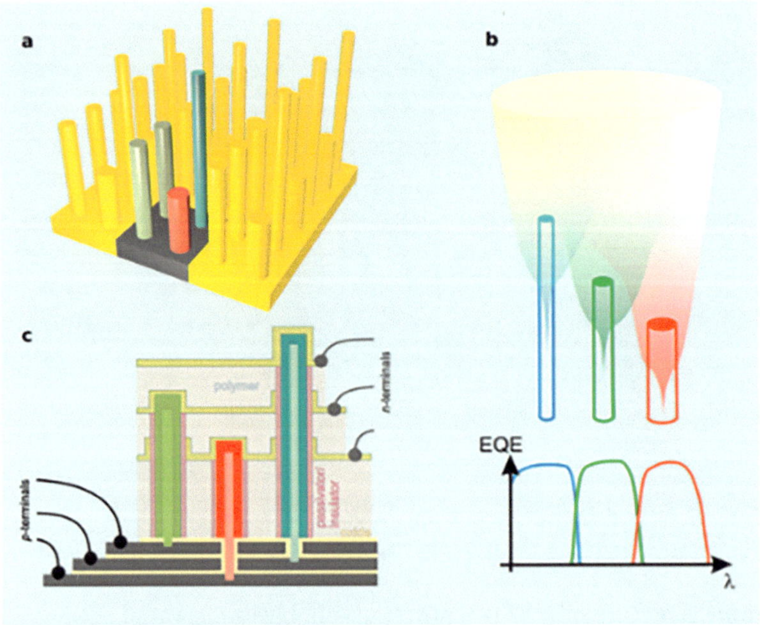

Figure 13.
(a) Design concept illustration of the triple-junction NW array on a Si substrate, (b) Working principle of the design, and (c) Contacting scheme of the multiterminal device [29].

polymer onto vertically aligned n-type GaAs NW arrays synthesized by MBE and reached an efficiency of 1.04% (2.6 sun) [106].

NWs can also be fabricated and designed as branch cells. Branched NWs solar cells [118–120] can also be referred to as nanotrees or nanoforests. These nanowires have a tunable 3D morphology, homo or heterogeneous junction, and interface electronic alignment represent a unique system for applications in energy conversion and storage devices. 3D branched nanowires have merits, including structural hierarchy, high surface areas, and direct electron transport pathways and it is an attractive recent research area on energy. Lundgren et al. [115] simulated a high absorption structure branched nanowire (BNW) (**Figure 14**). They found that BNW tree configurations achieved a maximum absorption of over 95% at 500 nm wavelength. There has been great progress in fabricating branched NWs [115]. Wang et al. [122] have reported the branched and hyperbranched NW synthesized by a multistep nanocluster-catalyzed VLS approach. They have demonstrated the growth of branched Si and GaN NWs with multi-generation branches.

Lightweight and flexible solar cells are necessarily important for designing high-efficiency solar cells [123]. Lightweight and flexibility are two of the desired properties, which can substantially reduce the facility weight, minimize the transportation cost, and cause the assumption of smart solar cells, such as integrating flexible cells into clothing. NWs provide unique merits in realizing these advanced functions, as they are going to be buried into polymers and then easily peeled away from the substrates. Han et al. [124] fabricated flexible GaAs NW solar cells with NWs lying horizontally and achieved high efficiency of 16% under atmosphere 1.5 global

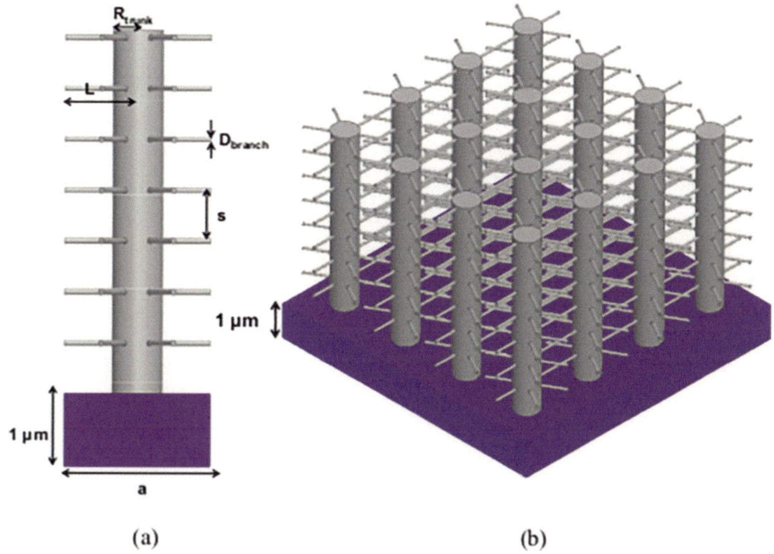

Figure 14.
(a) Dimensions of a single branched nanowire tree. (b) Array of branched NW trees [121].

illuminations. All the above discussed novel designs are very important for providing highly promising III–V NWs to greatly reduce the worth and boost the efficiency of solar cells, which may revolutionize the current solar cell technologies.

8. Recent efficiencies of III–V NWs solar cells

A solar cell is characterized by parameters, such as filling factor (FF), open-circuit voltage (V_{oc}), short-circuit current (J_{sc}), and power conversion efficiency (η). **Table 1** summarizes some of the fabricated III–V NWs solar cells since 2010. Values for each parameter and their growth mechanisms are also summarized. The main hindrance for commercializing III–V NWs solar cells is their low power conversion efficiency. Therefore, researchers around the world are trying to increase the efficiency of these materials by using novel designs, improving growth mechanisms, and device fabrication methods. The highest efficiency of III–V NW solar cells above 10% is reported by Holm et al. using GaAsP NWs with radial p-i-n junctions, which is 10.2% [133]. Next to this report, many III–V NWs with efficiency above 10% are reported. An efficiency of 19.6% using InP nanopillars is achieved by Son Ko et al. [145]. Krogstrup et al. even reported a high experimental efficiency of 40% using GaAs NWs [84].

Furthermore, the highest efficient III–V NWs large-area solar cells are rapidly developing [111]. The first large-area solar cell with high-efficiency higher than 10%, which is an InP NW array solar cell with an efficiency of 13.8%, is reported by Wallentin et al. [30]. Afterward, an efficiency of 11.1% with InP NW arrays is achieved by Cui et al. [136]. Then, an efficiency of 15.3% is achieved using GaAs NW arrays by Åberg et al. [76]. Ultimately, the III–V NW with the highest efficiency and world record is 17.8% reported by Dam et al. [32].

III–V nanowires	Growth methods	Substrates	catalysts	Geometry	FF (%)	V_{oc} (V)	η (%)	J_{sc} (mA/cm^2)	Ref.
InGaAs	SA-MOVPE	n-GaAs (111) B		axial p-i-n	72.1	0.544	7.14	18.2	[125]
InP	SA-MOCVD	InP (111) A		p-i-n single	74	0.69	2.81	5.52	[126]
InP	VLS	InP (111) B		axial p-i-n	68	0.61	7.6	18.2	[127]
GaAs	VLS	Si (111)	Ga	ZB crystal structure	40	0.45	4.1	22.8	[128]
InAs	MBE	p-Si (111)		p-n heterojunction	32	0.31	1.4	14	[129]
AlGaAs/GaAs heterojunctions	3-D optoelectronic simulation	[a]NI	NI	axial and radial with AlGaAs passivation	NI	NI	8.42	NI	[130]
GaAs	MBE	p-Si (111)	Au	axial p-n	NI	NI	6.3	NI	[131]
GaAs/InGaP/GaAs	MOVPE	p-GaAs (111) B	Au	core-multishell	52	0.5	4.7	18.1	[132]
GaAsP	MBE	p-Si (111)	Ga	core-shell p-i-n	77	0.9	10.2	14.7	[133]
$In_{0.3}Ga_{0.7}As$	MOVPE	p-Si (111)	free	n-InGaAs/p-Si	50	0.37	2.4	12	[134]
InN	MBE	n-Si (111)	free	axial p-i-n	30.24	0.13	0.51	12.91	[135]
InP	Top-down	InP		axial p-n	79.4	0.765	17.8	29.3	[32]
InP	MOVPE	p-InP (111) B	Au	axial p-i-n	72.4	0.906	13.8	24.6	[30]
InP	VLS	InP (111) B	Au	axial p-n junction	73	0.73	11.1	21	[136]
InP	SA-MOVPE	p-InP (111) A	free	ITO/p-InP heterojunction	68.2	0.436	7.37	24.8	[137]
InP	Top down	p-InP (100)		n-S-SAM/p-InP heterojunction	60	0.54	8.1	25	[36]
InP	SA-MOVPE	InP (111) A	free	NI	59.6	0.457	6.35	23.4	[138]
InP	SA-MOVPE	p-InP (111) A	free	radial p-i-n	58.5	0.674	4.23	11.1	[57]

DOI: http://dx.doi.org/10.5772/intechopen.105985

InAs hetero	MBE	p-Si (111)		n-InAs/p-Si	32	0.31	1.4	14	[129]
GaAs	MOVPE	p-GaAs (111) B	Au	axial p-i-n	79.2	0.906	15.3	21.3	[76]
GaAs	SSCVD	Non-crystalline	Au	Schottky contact	61	0.39	16	67	[124]
GaAs tandem	SAE-MOVPE	Si (111) n,p	free	axial n-i-p/n-p Si	57.8	0.956	11.4	20.64	[100]
GaAs	SAG-MOCVD	p-GaAs (111) B	free		63.65	0.565	7.8	21.08	[69]
GaAs	SA-MOVPE	n-GaAs (111) B	free	core-shell p-n	62	0.44	6.63	24.3	[139]
GaAs	MBE	p-Si (111)	Ga	core-shell p-i-n	52	0.43	40 (field concentration)	180 (field concentration)	[84]
GaAs/InGaP	SA-MOVPE	p-GaAs (111) B	free	Core-multishell p-n	65	0.5	4.01	12.7	[140]
GaAs	VLS	Si (111)	Ga	radial p-i-n	46.5	0.39	3.3	18.2	[141]
GaAs	SSCVD	Si/SiO_2	Au-Ga	Schottky contact	42	0.6	2.8	11	[142]
GaAs	SA-MOCVD	p-GaAs (111) B	free	core-shell p-n	37	0.39	2.54	17.6	[143]
GaAs hybrid	MOCVD	GaAs	free	GaAs/P3HT hybrid	43	0.2	1.44	18.6	[144]

[*] Power conversion efficiency was measured at 1 Sun, AM1.5G illumination; [a] NI represents not identified values.

Table 1.
Some of the III–V NWs studied since 2010 and their efficiency achievement.

9. Conclusion and outlook

In order to convert solar energy into electrical energy, harvesting solar energy is required and solar photovoltaic is the most promising device for this purpose. Despite the excess of sunlight reaching the earth's planet, the current percentage of solar energy is much smaller than both renewable and nonrenewable energies. This is due to low energy density, low efficiency, and relatively high-cost materials compared to other types of energy technologies. Therefore, novel materials that can enormously harvest sunlight are important and they are the current issues attracting research interest. Due to its unique properties from bulk materials, III–V NWs can be used as high-performance solar cells because of their attractive advantages, such as unique optical and electrical properties, direct band, and fewer solar light reflections. In constructing novel solar cells with high-efficiency and low-cost, the distinctive structure and advanced properties of NWs provide more freedom. Today's solar cell market is dominated by the thin film of Si, which has the lowest efficiency, but low cost. By combining the advantages of III–V NWs and Si by growing III–V NWs on Si substrate tandem solar cells, enormously improved performance of the solar cells can be achieved. By controlling the III–V NW morphology and its geometry with optimum diameter, period, and length it is possible to get high-efficiency solar cell materials. Furthermore, III–V NW solar cells can be designed as tandem solar cells, axial and radial tandem solar cells, multiterminal solar cells, inorganic nanowire/organic hybrid solar cells, branched solar cells, and flexible solar cells.

Future works can be focused on the optimum design that can overcome all the limitations of III–V NW solar cells in order to achieve high-performance and low-cost III–V NW-based solar cells. Thus, one of the best aspects of III–V NWs commercialization with high power conversion efficiency may be achieved by designing it in a way that it can absorb solar light enormously with reduction of materials used and low-cost substrates. According to our understanding, all the designs of III–V NWs that are mentioned in this review are beneficial for future commercialization; however, it is good to identify the one that is more attractive than the others by conducting research on each design. All types of design have their own advantages in case of reducing materials used for the fabrications of solar cells and cost reductions. So far, it is good if this area will be researched more, especially on the architecture of III–V NWs due to its infinite advantages. Despite the challenges of achieving high efficiency in these NWs, they are the hope of the next-generation solar cells due to their flexibility for designing it even in a multi-junction of different NWs, which can absorb the different wavelengths of solar light for harvesting huge solar light. Furthermore, to advance III–V NW-based solar cells toward possible commercialization the power conversion efficiency should be increased, for which the tandem architecture is highly interesting by growing on Si substrate, which is cost-effective.

Acknowledgements

We acknowledge Adama Science and Technology University and Oda Bultum University for their financial support.

Conflicts of interest

There is no conflict of interest.

DOI: http://dx.doi.org/10.5772/intechopen.105985

Author details

Fikadu Takele Geldasa[1,2]

1 Department of Applied Physics, Adama Science and Technology University, Adama, Ethiopia

2 School of Natural and Computational Sciences, Oda Bultum University, Chiro, Ethiopia

*Address all correspondence to: fikadutakele2008@gmail.com

References

[1] Wassie YT, Adaramola MS. Socio-economic and environmental impacts of rural electrification with solar photovoltaic systems: Evidence from southern Ethiopia. Energy for Sustainable Development. 2021;**60**:52-66. DOI: 10.1016/j.esd.2020.12.002

[2] U. Nations. Department of Economic and Social Affairs, Population Division (2019). World Population Prospects 2019, Volume I: Comprehensive Tables (ST/ESA/SER.A/426)

[3] Hasanuzzaman M, Islam MA, Rahim NA, Yanping Y. Energy demand. Energy for Sustainable Development. Elsevier; 2020. pp. 41-87

[4] Edenhofer O et al. Renewable energy sources and climate change mitigation: special report of the Intergovernmental Panel on Climate Change. Choice Reviews. 2012;**49**:49-6309-49-6309. DOI: 10.5860/CHOICE.49-6309

[5] Owusu PA, Asumadu-Sarkodie S. A review of renewable energy sources, sustainability issues and climate change mitigation. Cogent Engineering. 2016;**3**: 1-14. DOI: 10.1080/23311916.2016. 1167990

[6] IRENA. Future of Solar Photovoltaic: Deployment, investment, technology, grid integration and socio-economic aspects (A Global Energy Transformation: paper), International Renewable Energy Agency, Abu Dhabi. 2019

[7] Tsao J, Lewis N, Crabtree G. Solar FAQs. US Dep. Energy; 2006. pp. 1-24. Available from: http://www.sandia.gov/~jytsao/SolarFAQs.pdf

[8] Crabtree GW, Lewis NS. Solar energy conversion. Physics Today. 2007;**60**:37-42. DOI: 10.1063/1.2718755

[9] Baldwin SF. Renewable energy: Progress and prospects. Physics Today. 2002;**55**:62-67. DOI: 10.1063/1.1480784

[10] Paniyil P et al. Photovoltaics- and battery-based power network as sustainable source of electric power. Energies. 2020;**13**:1-22. DOI: 10.3390/en13195048

[11] Vasiliev M, Nur-E-Alam M, Alameh K. Recent developments in solar energy-harvesting technologies for building integration and distributed energy generation. Energies. 2019;**12**:1-23. DOI: 10.3390/en12061080

[12] Karakaya E, Sriwannawit P. Barriers to the adoption of photovoltaic systems : The state of the art. Renewable and Sustainable Energy Reviews. 2015;**49**:60-66. DOI: 10.1016/j.rser.2015.04.058

[13] Walters J, Kaminsky J, Gottschamer L. A systems analysis of factors influencing household solar PV adoption in Santiago, Chile. Sustainability. 2018;**10**:1-17. DOI: 10.3390/su10041257

[14] Almosni S et al. Material challenges for solar cells in the twenty-first century: Directions in emerging technologies. Science and Technology of Advanced Materials. 2018;**19**:336-369. DOI: 10.1080/14686996.2018.1433439

[15] Mariotti N et al. Recent advances in eco-friendly and cost-effective materials towards sustainable dye-sensitized solar cells. Green Chemistry. 2020;**22**:7168-7218. DOI: 10.1039/d0gc01148g

[16] Tiwari S, Tiwari T, Carter SA, Scott JC, Yakhmi JV. Advances in polymer-based photovoltaic cells: Review of pioneering materials, design, and device

physics. In: Handbook of Ecomaterials. Vol. 2. Cham: Springer International Publishing; 2019. pp. 1055-1101

[17] Wilson GM et al. The 2020 photovoltaic technologies roadmap. Journal of Physics D: Applied Physics. 2020;**53**:1-47. DOI: 10.1088/1361-6463/ab9c6a

[18] Lapierre RR et al. III-V nanowire photovoltaics: Review of design for high efficiency. Physica Status Solidi (RRL)– Rapid Research Letters. 2013;**7**:815-830. DOI: 10.1002/pssr.201307109

[19] Green MA. Third generation photovoltaics: Ultra-high conversion efficiency at low cost. Progress in Photovoltaics: Research and Applications. 2001;**9**:123-135. DOI: 10.1002/pip.360

[20] Shockley W, Queisser HJ. Detailed balance limit of efficiency of p-n junction solar cells. Journal of Applied Physics. 1961;**32**:510-519. DOI: 10.1063/1.1736034

[21] Cotal H et al. III-V multijunction solar cells for concentrating photovoltaics. Energy & Environmental Science. 2009;**2**:174-192. DOI: 10.1039/b809257e

[22] Zhang Y, Liu H. Nanowires for high-efficiency, low-cost solar photovoltaics. Crystals. 2019;**9**:1-24. DOI: 10.3390/cryst9020087

[23] Capper P, Irvine S, Joyce T. Epitaxial Crystal Growth: Methods and Materials. Springer, Springer Handbook of Electronic and Photonic Materials; 2017. pp. 309-341. DOI: 10.1007/978-3-319-48933-9

[24] Li J et al. A brief review of high efficiency III-V solar cells for space application. Frontiers of Physics. 2021;**8**:1-15. DOI: 10.3389/fphy.2020.631925

[25] Afzaal M, O'Brien P. Recent developments in II-VI and III-VI semiconductors and their applications in solar cells. Journal of Materials Chemistry. 2006;**16**:1597-1602. DOI: 10.1039/b512182e

[26] Mokkapati S, Jagadish C. Review on photonic properties of nanowires for photovoltaics. Optics Express. 2016;**24**. DOI: 10.1364/oe.24.017345

[27] Mohseni PK et al. Monolithic III-V nanowire solar cells on graphene via direct van der Waals epitaxy. Advanced Materials. 2014;**26**:3755-3760. DOI: 10.1002/adma.201305909

[28] Anttu N. Shockley-queisser detailed balance efficiency limit for nanowire solar cells. ACS Photonics. 2015;**2**: 446-453. DOI: 10.1021/ph5004835

[29] Dorodnyy A, Alarcon-lladó E, Shklover V, Hafner C. Efficient multi-terminal spectrum splitting via a nanowire array solar cell. ACS Photonics. 2015;**2**:1284-1288. DOI: 10.1021/acsphotonics.5b00222

[30] Wallentin J, Anttu N, Asoli D, Huffman M, Åberg I, Magnusson MH, et al. InP Nanowire Array Solar Cells Achieving 13.8% Efficiency by Exceeding the Ray Optics Limit. Nature. 2013;**353**:1-5. DOI: 10.1038/353737a0

[31] Li Z, Tan HH, Jagadish C, Fu L. III–V semiconductor single nanowire solar cells: A review. Advanced Materials Technologies. 2018;**3**:1-12. DOI: 10.1002/admt.201800005

[32] Van Dam D et al. High-efficiency nanowire solar cells with Omnidirectionally enhanced absorption due to self-aligned indium-tin-oxide Mie Scatterers. ACS Nano. 2016;**10**:11414-11419. DOI: 10.1021/acsnano.6b06874

[33] Hwang I, Um HD, Kim BS, Wober M, Seo K. Flexible crystalline silicon radial junction photovoltaics with vertically

aligned tapered microwires. Energy & Environmental Science. 2018;**11**:641-647. DOI: 10.1039/c7ee03340k

[34] Garnett EC, Brongersma ML, Cui Y, McGehee MD. Nanowire solar cells. Annual Review of Materials Research. 2011;**41**:269-295. DOI: 10.1146/annurev-matsci-062910-100434

[35] Za'Bah NF, Kwa KSK, Bowen L, Mendis B, O'Neill A. Top-down fabrication of single crystal silicon nanowire using optical lithography. Journal of Applied Physics. 2012;**112**. DOI: 10.1063/1.4737463

[36] Cho K et al. Molecular monolayers for conformal, nanoscale doping of InP nanopillar photovoltaics. Applied Physics Letters. 2011;**98**:27-30. DOI: 10.1063/1.3585138

[37] Elbersen R, Vijselaar W, Tiggelaar RM, Gardeniers H, Huskens J. Fabrication and doping methods for silicon Nano- and Micropillar arrays for solar-cell applications: A review. Advanced Materials. 2015;**27**:6781-6796. DOI: 10.1002/adma.201502632

[38] Lu Y, Lal A. High-efficiency ordered silicon nano-conical-frustum array solar cells by self-powered parallel electron lithography. Nano Letters. 2010;**10**:4651-4656. DOI: 10.1021/nl102867a

[39] Naulleau P. Optical lithography. In: Comprehensive Nanoscience and Nanotechnology, 2nd ed.; Andrews DL, Lipson RH, Nann T, Eds.; Oxford, UK: Academic Press. 2019; pp. 387-398, DOI: 10.1016/B978-0-12-803581-8.10433-3

[40] Vieu C et al. Electron beam lithography: Resolution limits and applications. Applied Surface Science. 2000;**164**:111-117. DOI: 10.1016/S0169-4332(00)00352-4

[41] Yu HK, Lee JL. Growth mechanism of metal-oxide nanowires synthesized by electron beam evaporation: A self-catalytic vapor-liquid-solid process. Scientific Reports. 2014;**4**:1-8. DOI: 10.1038/srep06589

[42] Costner EA, Lin MW, Jen W, Willson CG. Nanoimprint lithography materials development for semiconductor device fabrication. Annual Review of Materials Research. 1999;**39**:155-180. DOI: 10.1146/annurev-matsci-082908-145336

[43] Schift H, Kristensen A. Nanoimprint Lithography. Springer, Berlin, Heidelberg. Springer Handbooks; 2017. pp. 113-142. DOI: 10.1007/978-3-662-54357-3_5

[44] Lan H. Large-area nanoimprint lithography and applications. Micro/Nanolithograph IntechOpen. 2017:44-68. DOI: 10.5772/intechopen.72860

[45] Dirdal CA, Jensen GU, Angelskår H, Vaagen Thrane PC, Gjessing J, Ordnung DA. Towards high-throughput large-area metalens fabrication using UV-nanoimprint lithography and Bosch deep reactive ion etching. Optics Express. 2020;**28**:15542-15561. DOI: 10.1364/oe.393328

[46] Altay MC. Growth of Ge nanowires by chemical vapor deposition at atmospheric pressure using readily available precursors GeO_2 and C_2H_5OH. JOM. 2020;**72**:4340-4345. DOI: 10.1007/s11837-020-04401-3

[47] Toros A et al. Reactive ion etching of single crystal diamond by inductively coupled plasma: State of the art and catalog of recipes. Diamond and Related Materials. 2020;**108**:1-17. DOI: 10.1016/j.diamond.2020.107839

[48] Pearton SJ, Shul RJ, Ren F. A review of dry etching of GaN and related materials. MRS Internet Journal of Nitride Semiconductor Research. 2000;**5**. DOI: 10.1557/S1092578300000119

[49] Wolfsteller A et al. Comparison of the top-down and bottom-up approach to fabricate nanowire-based silicon/germanium heterostructures. Thin Solid Films. 2010;**518**:2555-2561. DOI: 10.1016/j.tsf.2009.08.021

[50] Diedenhofen SL et al. Broad-band and omnidirectional antireflection coatings based on semiconductor nanorods. Advanced Materials. 2009;**21**:973-978. DOI: 10.1002/adma.200802767

[51] Puglisi RA et al. Chemical vapor deposition growth of silicon nanowires with diameter smaller than 5 nm. ACS Omega. 2019;**4**:17967-17971. DOI: 10.1021/acsomega.9b01488

[52] Persson AI et al. Surface diffusion effects on growth of nanowires by chemical beam epitaxy surface diffusion effects on growth of nanowires by chemical beam epitaxy. Journal of Applied Physics. 2007;**101**:1-6(0343131). DOI: 10.1063/1.2435800

[53] Nunez CG, De Cal AFB, Lopez N, Javier B, Carretero G. A novel growth method to improve the quality of GaAs nanowires grown by Ga-assisted chemical beam epitaxy. Nano Letters. 2018;**6**:3608-3615. DOI: 10.1021/acs.nanolett.8b00702

[54] Gordon EB, Stepanov ME, Kulish MI, Karabulin AV, Matyushenko VI, Khodos II. The nanowires growth by laser ablation of metals inside rotating superfluid helium. Laser Physics Letters. 2019;**16**: 1-6. DOI: 10.1088/1612-202X/aaf6a1

[55] Saddiqi N, Javed H, Islam M, Ghauri K. A review on synthesis of silicon nanowires by laser ablation. Chemistry and Materials Research. 2014;**6**:76-86

[56] Behroudj A, Geiger D, Strehle S. Epitaxial bottom-up growth of silicon nanowires on oxidized silicon by alloy-catalyzed gas-phase synthesis. Nano Letters. 2019;**11**:7895-7900. DOI: 10.1021/acs.nanolett.9b02950

[57] Fukui T, Yoshimura M, Nakai E, Tomioka K. Position-controlled III-V compound semiconductor nanowire solar cells by selective-area metal-organic vapor phase epitaxy. Ambio. 2012;**41**:119-124. DOI: 10.1007/s13280-012-0266-5

[58] Strudley T, Zehender T, Blejean C, Bakkers EPAM, Muskens OL. Mesoscopic light transport by very strong collective multiple scattering in nanowire mats. Nature Photonics. 2013;**7**:413-418. DOI: 10.1038/nphoton.2013.62

[59] Yoshimoto N, Matsuoka T, Sasaki T, Katsui A. Photoluminescence of InGaN films grown at high temperature by metalorganic vapor phase epitaxy. Applied Physics Letters. 1991;**59**: 2251-2253. DOI: 10.1063/1.106086

[60] Ikejiri K, Noborisaka J, Hara S, Motohisa J, Fukui T. Mechanism of catalyst-free growth of GaAs nanowires by selective area MOVPE. Journal of Crystal Growth. 2007;**298**:616-619. DOI: 10.1016/j.jcrysgro.2006.10.179

[61] Aseev P et al. Selectivity map for molecular beam epitaxy of advanced III-V quantum nanowire networks. Nano Letters. 2019;**19**:218-227. DOI: 10.1021/acs.nanolett.8b03733

[62] Jabeen F, Rubini S, Martelli F. Growth of III-V semiconductor nanowires by molecular beam epitaxy. Microelectronics Journal. 2009;**40**: 442-445. DOI: 10.1016/j.mejo.2008.06.001

[63] Mandl B et al. Growth mechanism of self-catalyzed group III-V nanowires. Nano Letters. 2010;**10**:4443-4449. DOI: 10.1021/nl1022699

[64] Noborisaka J, Motohisa J, Fukui T. Catalyst-free growth of GaAs nanowires by selective-area metalorganic vapor-phase epitaxy. Applied Physics Letters. 2005;**86**:1-3. DOI: 10.1063/1.1935038

[65] Sokolovskii AS, Robson MT, Lapierre RR, Dubrovskii VG. Modeling selective-area growth of InAsSb nanowires. Nanotechnology. 2019;**30**:1-7. DOI: 10.1088/1361-6528/ab1375

[66] Rackauskas S, Nasibulin AG. Nanowire growth without catalysts: Applications and mechanisms at the atomic scale. ACS Applied Nano Materials. 2020;**3**:7314-7324. DOI: 10.1021/acsanm.0c01179

[67] Rudolph D et al. Direct observation of a noncatalytic growth regime for GaAs nanowires. Nano Letters. 2011;**11**:3848-3854. DOI: 10.1021/nl2019382

[68] Zhang RQ, Lifshitz Y, Lee ST. Oxide-assisted growth of semiconducting nanowires. Advanced Materials. 2003;**15**:635-640. DOI: 10.1002/adma.200301641

[69] Yao M et al. GaAs nanowire array solar cells with axial p-i-n junctions. Nano Letters. 2014;**14**:3293-3303. DOI: 10.1021/nl500704r

[70] Huson JJ, Sheng T, Ogle E, Zhang H. Reaction intermediate-induced vapor–liquid–solid growth of silicon oxide nanowires. CrystEngComm. 2018;**20**:7256-7265. DOI: 10.1039/c8ce01115j

[71] Maliakkal CB et al. In situ analysis of catalyst composition during gold catalyzed GaAs nanowire growth. Nature Communications. 2019;**10**:1-9. DOI: 10.1038/s41467-019-12437-6

[72] Vijayakumar Sheela H, Madhusudhanan V, Krishnan G. Substrate-independent and catalyst-free synthesis of magnesium nanowires. Nanoscale Advances. 2019;**1**:1754-1762. DOI: 10.1039/c9na00072k

[73] Zhang C, Miao X, Mohseni PK, Choi W, Li X. Site-controlled VLS growth of planar nanowires: Yield and mechanism. Nano Letters. 2014;**14**:6836-6841. DOI: 10.1021/nl502525z

[74] Noor Mohammad S. Why self-catalyzed nanowires are most suitable for large-scale hierarchical integrated designs of nanowire nanoelectronics. Journal of Applied Physics. 2011;**110**. DOI: 10.1063/1.3624585

[75] Persson AI, Larsson MW, Stenström S, Ohlsson BJ, Samuelson L, Wallenberg LR. Solid-phase diffusion mechanism for GaAs nanowire growth. Nature Materials. 2004;**3**:677-681. DOI: 10.1038/nmat1220

[76] Åberg I, Vescovi G, et al. A GaAs nanowire Array solar cell with 15.3% efficiency at 1 Sun. IEEE Journal of Photovoltaics. 2016;**6**:185-190. DOI: 10.1109/JPHOTOV.2015.2484967

[77] Goktas NI, Wilson P, Ghukasyan A, Wagner D, McNamee S, LaPierre RR. Nanowires for energy: A review. Applied Physics Reviews. 2018;**5**:1-21. DOI: 10.1063/1.5054842

[78] Ayvazyan G, Barseghyan R, Minasyan S. Optimization of surface reflectance for silicon solar cells. E3S Web Conferences. 2018;**69**:2-5. DOI: 10.1051/e3sconf/20186901008

[79] Sharma R. Effect of obliquity of incident light on the performance of silicon solar cells. Heliyon. 2019;**5**:2-4. DOI: 10.1016/j.heliyon.2019.e01965

[80] Brongersma ML, Cui Y, Fan S. Light management for photovoltaics

using high-index nanostructures. Nature Materials. 2014;**13**:451-460. DOI: 10.1038/nmat3921

[81] Mavrokefalos A, Han SE, Yerci S, Branham MS, Chen G. Efficient light trapping in inverted nanopyramid thin crystalline silicon membranes for solar cell applications. Nano Letters. 2012;**12**:2792-2796. DOI: 10.1021/nl2045777

[82] Wu D, Tang X, Wang K, He Z, Li X. An efficient and effective design of InP nanowires for maximal solar energy harvesting. Nanoscale Research Letters. 2017;**12**:1-10. DOI: 10.1186/s11671-017-2354-8

[83] Wen L, Zhao Z, Li X, Shen Y, Guo H, Wang Y. Theoretical analysis and modeling of light trapping in high efficicency GaAs nanowire array solar cells. Applied Physics Letters. 2011;**99**:2009-2012. DOI: 10.1063/1.3647847

[84] Krogstrup P et al. Single-nanowire solar cells beyond the Shockley-Queisser limit. Nature Photonics. 2013;7:306-310. DOI: 10.1038/nphoton.2013.32

[85] Cao L et al. Semiconductor nanowire optical antenna solar absorbers. Nano Letters. 2010;**10**:439-445. DOI: 10.1021/nl9036627

[86] Callahan DM, Munday JN, Atwater HA. Solar cell light trapping beyond the ray optic limit. Nano Letters. 2012;**12**:214-218. DOI: 10.1021/nl203351k

[87] Wang W, Wu S, Reinhardt K, Lu Y, Chen S. Broadband light absorption enhancement in thin-film silicon solar cells. Nano Letters. 2010;**10**:2012-2018. DOI: 10.1021/nl904057p

[88] Zhang X, Sun XH, Di Jiang L. Absorption enhancement using nanoneedle array for solar cell. Applied Physics Letters. 2013;**103**:1-5. DOI: 10.1063/1.4832216

[89] Jeong S, Wang S, Cui Y. Nanoscale photon management in silicon solar cells. Journal of Vacuum Science and Technology A. 2012;**30**:1-11. DOI: 10.1116/1.4759260

[90] Hu Y, Lapierre RR, Li M, Chen K, He JJ. Optical characteristics of GaAs nanowire solar cells. Journal of Applied Physics. 2012;**112**:1-6. DOI: 10.1063/1.4764927

[91] Raj V, Tan HH, Chennupati J. Axial vs radial junction nanowire solar cell. Asian Journal of Physics. 2019;**28**:719-746

[92] Hodes G, Kamat PV. Understanding the implication of carrier diffusion length in photovoltaic cells. Journal of Physical Chemistry Letters. 2015;**6**:4090-4092. DOI: 10.1021/acs.jpclett.5b02052

[93] David A. Long-range carrier diffusion in (In , Ga) N quantum wells and implications from fundamentals to devices. Physical Review Applied. 2021;**15**:1-9. DOI: 10.1103/physrevapplied.15.054015

[94] Kayes BM, Atwater HA, Lewis NS. Comparison of the device physics principles of planar and radial p-n junction nanorod solar cells. Journal of Applied Physics. 2005;**97**:1-11. DOI: 10.1063/1.1901835

[95] Thiyagu S, Pei Z, Jhong MS. Amorphous silicon nanocone array solar cell. Nanoscale Research Letters. 2012;**7**:1-6. DOI: 10.1186/1556-276X-7-1

[96] Zhang Y et al. Growth and fabrication of high-quality single nanowire devices with radial p-i-n junctions. Small. 2019;**15**:1-7. DOI: 10.1002/smll.201803684

[97] Chen Y, Pistol M, Anttu N. Design for strong absorption in a nanowire array tandem solar cell. Scientific Reports. 2016;**6**:1-8. DOI: 10.1038/srep32349

[98] Hu Y, Li M, He JJ, Lapierre RR. Current matching and efficiency optimization in a two-junction nanowire-on-silicon solar cell. Nanotechnology. 2013;**24**:1-5. DOI: 10.1088/0957-4484/24/6/065402

[99] Connolly JP, Mencaraglia D, Renard C, Bouchier D. Designing III-V multijunction solar cells on silicon. Progress in Photovoltaics: Research and Applications. 2014;**22**:810-820. DOI: 10.1002/pip.2463

[100] Yao M et al. Tandem solar cells using GaAs nanowires on Si: Design, fabrication, and observation of voltage addition. Nano Letters. 2015;**15**:7217-7224. DOI: 10.1021/acs.nanolett.5b03890

[101] Kempa TJ, Tian B, Kim DR, Hu J, Zheng X, Lieber CM. Single and tandem axial p-i-n nanowire photovoltaic devices. Nano Letters. 2008;**8**:3456-3460. DOI: 10.1021/nl8023438

[102] Huang N, Lin C, Povinelli ML, Huang N, Lin C, Povinelli ML. Limiting efficiencies of tandem solar cells consisting of III-V nanowire arrays on silicon limiting efficiencies of tandem solar cells consisting of III-V nanowire arrays on silicon. Journal of Applied Physics. 2013;**112**. DOI: 10.1063/1.4754317

[103] Zhong Z et al. Efficiency enhancement of axial junction InP single nanowire solar cells by dielectric coating. Nano Energy. 2016;**28**:106-114. DOI: 10.1016/j.nanoen.2016.08.032

[104] Wan J, Song T, Flox C, Yang J, Yang Q, Han X. Advanced nanomaterials for energy-related applications. Journal of Nanomaterials. 2015;**2015**:1-2. DOI: 10.1155/2015/564097

[105] Uccelli E, Arbiol J, Morante JR, Fontcuberta A, Morral I. InAs quantum dot arrays decorating the facets of GaAs nanowires. ACS Nano. 2010;**4**:5985-5993. DOI: 10.1021/nn101604k

[106] Boulanger JP, Lapierre RR. Polytype formation in GaAs/GaP axial nanowire heterostructures. Journal of Crystal Growth. 2011;**332**:21-26. DOI: 10.1016/j.jcrysgro.2011.07.021

[107] Ercolani D et al. InAs/InSb nanowire heterostructures grown by chemical beam epitaxy. Nanotechnology. 2009;**20**:1-6. DOI: 10.1088/0957-4484/20/50/505605

[108] Haapamaki CM, Lapierre RR. Mechanisms of molecular beam epitaxy growth in InAs / InP nanowire heterostructures. Nanotechnology. 2011;**22**. DOI: 10.1088/0957-4484/22/33/335602

[109] Lange H, Kelley DF. Spectroscopic effects of lattice strain in InP/ZnSe and InP/ZnS nanocrystals. Journal of Physical Chemistry C. 2020;**41**:22839-22844. DOI: 10.1021/acs.jpcc.0c07145

[110] Adachi MM, Anantram MP, Karim KS. Core-shell silicon nanowire solar cells. Scientific Reports. 2013;**3**:2-7. DOI: 10.1038/srep01546

[111] Nazarenko MV et al. Elastic energy relaxation and critical thickness for plastic deformation in the core-shell InGaAs/GaAs nanopillars. Journal of Applied Physics. 2013;**113**:(1-6)104311. DOI: 10.1063/1.4795168

[112] Wang S, Yan X, Zhang X, Li J, Ren X. Axially connected nanowire core-shell p-n junctions : A composite structure for high-efficiency solar cells.

Nanoscale Research Letters. 2015;**10**:0-6. DOI: 10.1186/s11671-015-0744-3

[113] Heurlin M et al. Axial InP nanowire tandem junction grown on a silicon substrate. Nano Letters. 2011;**11**:2028-2031. DOI: 10.1021/nl2004219

[114] Dorodnyy A, Lladó EA, Shklover V, Hafner C, Morral AF, Leuthold J. Efficient multiterminal Spectrum splitting via a nanowire Array solar cell. ACS Photonics. 2015;**2**:1284-1288. DOI: 10.1021/acsphotonics.5b00222

[115] Lundgren C, Lopez R, Redwing J, Melde K. FDTD modeling of solar energy absorption in silicon branched nanowires. Optics Express. 2013;**21**:A392-A400. DOI: 10.1364/OE.21.00A392

[116] Liu K, Qu S, Zhang X, Tan F, Wang Z. Improved photovoltaic performance of silicon nanowire / organic hybrid solar cells by incorporating silver nanoparticles. Nanoscale Research Letters. 2013;**8**:1-7. DOI: 10.1186/1556-276X-8-88

[117] Ren S, Chang L, Lim S, Zhao J, Smith M, Zhao N. Inorganic-organic hybrid solar cell : Bridging quantum dots to conjugated polymer nanowires. Nano Letters. 2011;**11**:3998-4002. DOI: 10.1021/nl202435t

[118] Cheng HM, Chiu WH, Lee CH, Tsai SY, Hsieh WF. Formation of branched ZnO nanowires from solvothermal method and dye-sensitized solar cells applications. Journal of Physical Chemistry C. 2008;**112**:16359-16364. DOI: 10.1021/jp805239k

[119] Liu J, Tao H, Cao Y, Ackermann J. Multi-branched cdse nanocrystals stabilized by weak ligand for hybrid solar cell application. Journal of Nanoscience and Nanotechnology. 2014;**14**:2836-2841. DOI: 10.1166/jnn.2014.8612

[120] Bierman MJ, Jin S. Potential applications of hierarchical branching nanowires in solar energy conversion. Energy & Environmental Science. 2009;**2**:1050-1059. DOI: 10.1039/b912095e

[121] Um H, Choi D, Choi A, Seo JH, Seo K. Embedded metal electrode for organic–inorganic hybrid nanowire solar cells. ACS Nano. 2017;**11**:6218-6224. DOI: 10.1021/acsnano.7b02322

[122] Wang D, Qian F, Yang C, Zhong Z, Lieber CM. Rational growth of branched and hyperbranched nanowire structures. Nano Letters. 2004;**4**:871-874. DOI: 10.1021/nl049728u

[123] Law DC et al. Lightweight, flexible, high-efficiency III-V Multijunction cells. IEEE. 4th World Conference on Photovoltaic Energy Conference. 2006. pp. 1879-1882. DOI: 10.1109/WCPEC.2006.279862

[124] Han N et al. High-performance GaAs nanowire solar cells for flexible and transparent photovoltaics. ACS Applied Materials & Interfaces. 2015;**7**:20454-20459. DOI: 10.1021/acsami.5b06452

[125] Nakai E, Chen M, Yoshimura M, Tomioka K, Fukui T. InGaAs axial-junction nanowire-array solar cells. Japanese Journal of Applied Physics. 2015;**54**:(1-4)47015201. DOI: 10.7567/JJAP.54.015201

[126] Zhong ZQ, Li ZY, Fu L, Gao Q, Li Z, Peng K, et al. InP single nanowire solar cells. Light, Energy and the Environment. 2015. DOI: 10.1364/pv.2015.ptu3b.3

[127] Cavalli A, Cui Y, Kölling S, Verheijen MA, Plissard SR, Wang J, et al. Influence of growth conditions on the performance of InP nanowire

solar cells. Nanotechnology. 2016;**27**:1-9. DOI: 10.1088/0957-4484/27/45/454003

[128] Dastjerdi MHT, Boulanger JP, Kuyanov P, Aagesen M, Lapierre RR. Methods of Ga droplet consumption for improved GaAs nanowire solar cell efficiency. Nanotechnology. 2016;**27**:1-7. DOI: 10.1088/0957-4484/27/47/475403

[129] Mallorquí AD et al. Characterization and analysis of InAs/p-Si heterojunction nanowire-based solar cell. Journal of Physics D: Applied Physics. 2014;**47**:1-8. DOI: 10.1088/0022-3727/47/39/394017

[130] Wu Y, Yan X, Wei W, Zhang J, Zhang X, Ren X. Optimization of GaAs nanowire pin junction array solar cells by using AlGaAs/GaAs heterojunctions. Nanoscale Research Letters. 2018;**13**:1-7. DOI: 10.1186/s11671-018-2503-8

[131] Alekseev PA, Sharov VA, Borodin BR, Dunaevskiy MS, Reznik RR, Cirlin GE. Effect of the uniaxial compression on the GaAs nanowire solar cell. Micromachines. 2020;**11**:1-13. DOI: 10.3390/mi11060581

[132] Gutsche C et al. N-GaAs/InGaP/p-GaAs core-multishell nanowire diodes for efficient light-to-current conversion. Advanced Functional Materials. 2012;**22**:929-936. DOI: 10.1002/adfm.201101759

[133] Holm JV, Jørgensen HI, Krogstrup P, Nygård J, Liu H, Aagesen M. Surface-passivated GaAsP single-nanowire solar cells exceeding 10% efficiency grown on silicon. Nature Communications. 2013;**4**:1-5. DOI: 10.1038/ncomms2510

[134] Shin JC et al. InxGa1-xAs nanowires on silicon: One-dimensional heterogeneous epitaxy, bandgap engineering, and photovoltaics. Nano Letters. 2011;**11**:4831-4838. DOI: 10.1021/nl202676b

[135] Nguyen HPT, Chang YL, Shih I, Mi Z. InN p-i-n nanowire solar cells on Si. IEEE Journal of Selected Topics in Quantum Electronics. 2011;**17**:1062-1069. DOI: 10.1109/JSTQE.2010.2082505

[136] Cui Y et al. Efficiency enhancement of InP nanowire solar cells by surface cleaning. Nano Letters. 2013;**13**: 4113-4117. DOI: 10.1021/nl4016182

[137] Yoshimura M, Nakai E, Tomioka K, Fukui T. Indium tin oxide and indium phosphide heterojunction nanowire array solar cells. Applied Physics Letters. 2013;**103**:24-27. DOI: 10.1063/1.4847355

[138] Yoshimura M, Nakai E, Tomioka K, Fukui T. Indium phosphide core—Shell nanowire array solar cells with lattice-mismatched window layer. Applied Physics Express. 2013;**6**:052301(1-4). DOI: 10.7567/APEX.6.052301

[139] Mariani G, Scofield AC, Hung CH, Huffaker DL. GaAs nanopillar-array solar cells employing in situ surface passivation. Nature Communications. 2013;**4**:1-7. DOI: 10.1038/ncomms2509

[140] Nakai E, Yoshimura M, Tomioka K, Fukui T. GaAs / InGaP Core—Multishell nanowire-Array- based solar cells. Japanese Journal of Applied Physics. 2013;**52**:1-4. DOI: 10.7567/JJAP.52.055002

[141] Boulanger JP et al. Characterization of a Ga-assisted GaAs nanowire Array solar cell on Si substrate. IEEE Journal of Photovoltaics. 2016;**6**:661-667. DOI: 10.1109/JPHOTOV.2016.2537547

[142] Han N et al. GaAs nanowire Schottky barrier photovoltaics utilizing Au-Ga alloy catalytic tips. Applied Physics Letters. 2012;**101**:(1-4)013105. DOI: 10.1063/1.4727907

[143] Mariani G, Wong PS, Katzenmeyer AM, Léonard F, Shapiro J,

Huffaker DL. Patterned radial GaAs nanopillar solar cells. Nano Letters. 2011;**11**:2490-2494. DOI: 10.1021/nl200965j

[144] Mariani G et al. Hybrid conjugated polymer solar cells using patterned GaAs nanopillars. Applied Physics Letters. 2010;**97**:3-6. DOI: 10.1063/1.3459961

[145] Ko WS, Tran TTD, Bhattacharya I, Ng KW, Sun H, Chang-hasnain C. Illumination angle insensitive single indium phosphide tapered Nanopillar solar cell. Nano Letters. 2015;**15**:4961-4967. DOI: 10.1021/acs.nanolett.5b00756

Chapter 6

What is the Profitability of a Photovoltaic Installation in France for an Individual?

Quentin Lagarde, Bruno Beillard, Serge Mazen and Julien Leylavergne

Abstract

The energy transition will require the use of renewable energy resources that will allow us to reach our decarbonization objectives. In addition to states and institutions, individuals have an important role to play in this transition, particularly with the installation of photovoltaic panels. But for individuals to use this source of energy, they must be guaranteed a financial return that encourages them to take the plunge. Several companies, in a commercial approach, guarantee returns on investment after 2–3 years for any installation. But is the financial profitability always guaranteed? In France, several types of photovoltaic installations are possible, total resale and self-consumption. For each one, the profitability will vary and depends on many parameters such as the initial and daily investment, the irradiation, the electricity buy-back price, and the consumption. This paper explains the calculation methodology for both typologies and shows that currently, in most cases, a self-consumption installation is more profitable than a full resale installation but is far from obtaining the returns on investment predicted by the commercials. If self-consumption is more profitable today, this is due to the fact that the investment (initial or annual) is less important, and that the price of electricity becomes more and more expensive while at the same time the price of resale decreases from year to year.

Keywords: photovoltaic panels, profitability, total resale, self-consumption, individuals

1. Introduction

The price of electricity increases every year. Moreover, with the global geopolitical context and the French energy context (planned partial replacement of the nuclear part by renewable energies), the cost of energy is likely to increase further. In addition, the massive arrival of connected objects (IoT), the increase of the electric car fleet, the democratization of telecommuting which will be accompanied by an increase in residential consumption. Therefore, the French are looking for solutions to reduce their energy bill. One of the possibilities is the installation of devices producing

IntechOpen

electricity from renewable energy, such as photovoltaic solar panels, which appear today as the most promising option for 31% of the French [1].

Indeed, the installation of photovoltaic panels in France is growing. At the end of 2019, 4,54,934 photovoltaic installations were recorded, corresponding to an installed capacity of 9904 MW, including 28,683 new installations in 2019 alone (966 additional MW) [2].

Of the 37 million customers in France registered by Enedis, 1.23% have a photovoltaic installation. Thus, the capacity for panel installation is still immense.

In 2019, 87% of installations were below 9 kWp (57% <3 kWp, 30% >3 kWp, and < 9 kWp) proving that the French prefer house-scale installations, and that the notion of collective production has not yet fully arrived in France [3].

Moreover, self-consumption has been developing more than the total resale since 2015. Between 2018 and 2019, 85% of installations were for self-consumption with surplus sale (158 MW) or without sale and without injection (36 MW). But all the self-consumption finally represents only 1.95% of the total French solar park.

Nevertheless, 1 out of 2 French people find that the investment is too expensive and becomes a brake to the installation of panels.

But really, is the installation of photovoltaic panels advantageous in all situations for the individual?

2. The different types of installations

The first difficulty for French people who want to invest in photovoltaic panels is to choose the type of their future installation. There are currently two main families.

The first is the total resale also called total injection (**Figure 1**) [4]. All the energy produced is resold to a supplier and is redistributed on the national grid. The supply of the house is classically done by the network. Here the individual will still be subject to

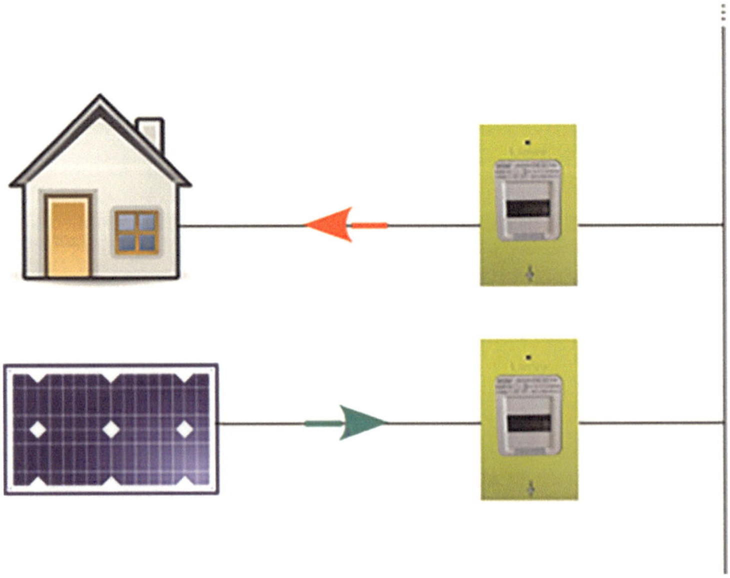

Figure 1.
Total resale.

the inflation of electricity. This type of installation is finally a simple financial investment in renewable energy.

The second possibility is self-consumption, which has been widely developed in 2015 (**Figure 2**) [5]. The energy produced is primarily consumed directly by the individual. Some variants exist for the management of the surplus energy. It can either be injected into the grid and bought back by the supplier or not sold and then lost or stored in batteries or virtually. The self-consumption allows direct reduction of the electricity bill of a private individual.

In the continuation of this article, the profitability of installation in total resale and in self-consumption with surplus will be compared.

3. Methods of calculation

The profitability of a photovoltaic installation for the total resale follows Eq. (1):

$$Gain(n,i,o) = Ep_{average}(n,i,o) \times Tvt - ID - \sum_{n=1}^{20}((IA)(n-1) \times [1 + infla2]) \qquad (1)$$

Where

n: the number of years, which is 20 years (time of the energy buy-back contract)

i: tilt of the panels

o: orientation of the panels

$Ep_{average}$: average energy produced over a year, starting from an average irradiation

Tvt: buy-back price of electricity in €/kWh

ID: initial investment in €.

IA: annual investment in €/year

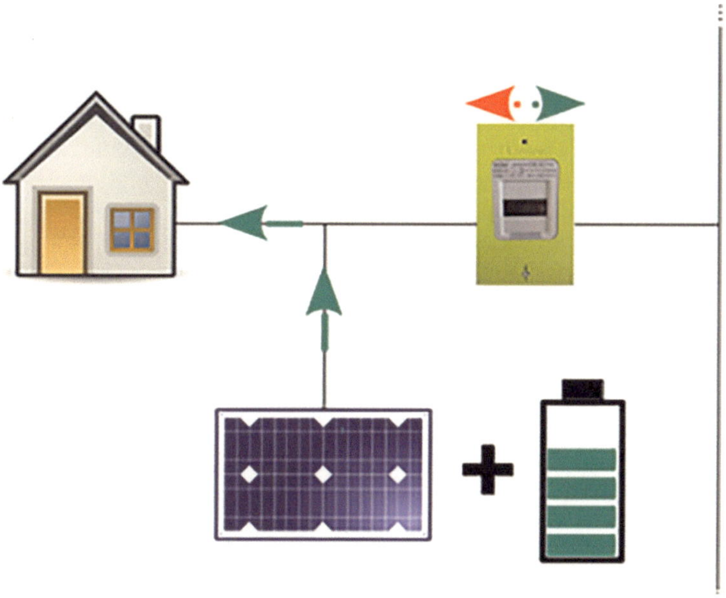

Figure 2.
Self-consumption.

Infla2: inflation of the annual investment price

The profitability of a photovoltaic installation for self-consumption differs slightly and follows Eq. (2):

$$
\begin{aligned}
Gain(n,i,o) = & \sum_{n=1}^{20}\left(\tau_{conso} \times Ep_{average}(n-1,i,o) \times THP(n-1) \times [1+infla]\right) \\
& + \sum_{n=1}^{20}\left((1-\tau_{conso}) \times Ep_{average}(n-1,i,o) \times Tvt\right) \\
& - \sum_{n=1}^{20}((IA)(n-1) \times [1+infla2]) - ID
\end{aligned}
\tag{2}
$$

Where

n: the number of years, which is 20 years (time of the energy buy-back contract)

i: tilt of the panels

o: orientation of the panels

$Ep_{average}$: average energy produced over a year starting from an average irradiation

τ_{self}: self-consumption rate

THP: peak hour tariff in €/kWh

Tvt: feed-in tariff in €/kWh

ID: initial investment in €.

IA: annual investment in €/year

Infla: inflation of electricity

Infla2: Inflation of the annual investment price

The different parameters of the profitability calculation are explained in the following sections.

3.1 Initial investment ID

The material selected for the realization of a photovoltaic project is essential. It must have the best quality/price ratio, guaranteeing an optimal yield for at least 20 years.

To sell the electricity to a supplier, it is mandatory that the installation is carried out by a certified company.

The installation will vary depending on the installers, the material, and the geographical area and follows a price range according to the following equations (see **Figure 3**):

Low range: 2.25-0.06*kWp €/Wp

High range: 3.29-0.13*kWp €/Wp

The price of the installation includes the solar panels, the inverters (inverters or micro-inverters), the decoupling relays, the cables, the protection boxes, the fixing kits, the transport, the labour, the security of the building site and the administrative procedures [6].

For example, an installation of 3 kWp of panels will cost the private individual with the installation by an approved installer and the price of the material between 6200 and 8700 €.

For the total resale, the connection fees of Enedis (Electricity Distribution System Operator) will be added. Prices vary according to the place of residence and the difficulty of connection. With a Linky meter, it is between 400 and 800 € and with the old meters between 1000 and 1400 €.

On the other hand, for self-consumption, mainly with resale of surplus, state bonuses are possible, ranging from 380 to 280 €/kWp depending on the power installed.

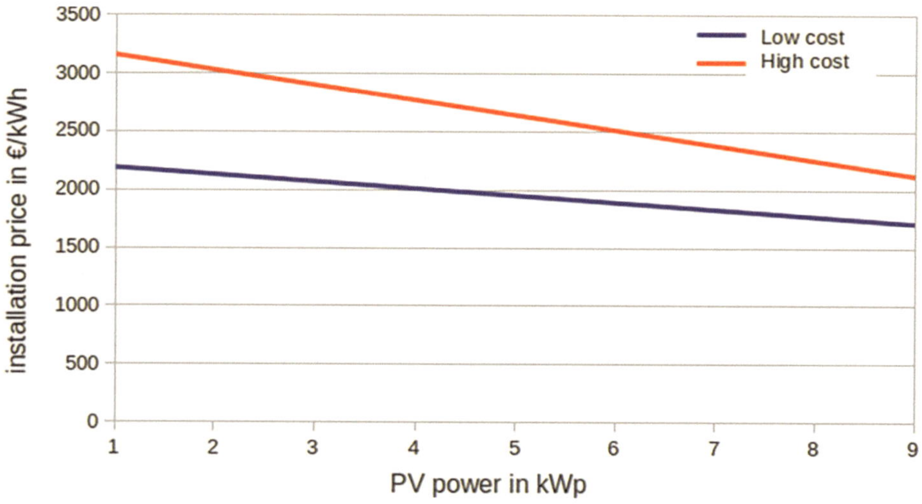

Figure 3.
High and low range of a photovoltaic installation in €/Wp according to the installed power in kWp in France.

First semester 2022	Total resale	Self-consumption with resale	Self-consumption without resale
Installation price (€/kWp)	between 2255–60*kWp and 3290–130*kWp	Between 2255–60*kWp and 3290–130*kWp	between 2255–60*kWp and 3290–130*kWp
Network connection (€)	between 400 and 1400	0	0
State aid (€/kWp)	0	< 3 kWp: 380 < 9 kWp: 280	0

Table 1.
The different sources of initial investment to be considered for the profitability calculation.

All the initial investments are summarized in **Table 1**.

The price range of the initial investment as a function of the installed power, between the 2 blue curves (maximum and minimum investment possible according to **Table 1**) for the total resale or between the 2 red curves for self-consumption, is presented in **Figure 4**. The initial investment is more advantageous in self-consumption because there is no need for connection to the grid and there is also a premium from the state.

3.2 Annual investment IA

In addition to the initial investment, there are annual charges.

There is the "tariff for the use of public electricity networks," or TURPE in French, created in 2000 to remunerate the electricity transmission network, Enedis and the local distribution companies [7]. It is currently 35.45 € including taxes/year for the total resale and 8.86 € including taxes/year for self-consumption with surplus resale (**Figure 5**). Since 2017, TURPE has changed very little over time and is cheaper for self-consumption.

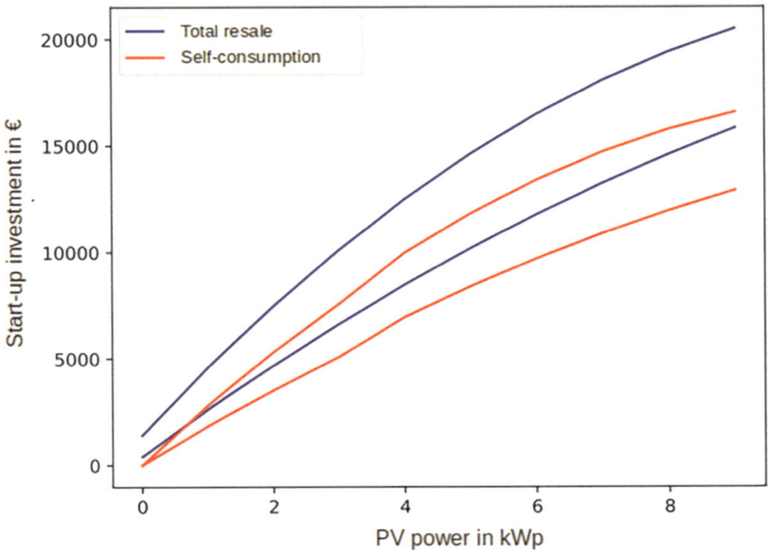

Figure 4.
Extremum for a starting investment ID for the total resale (blue curves) and self-consumption (red curves).

The maintenance of the installation (panels, inverters) can also be a source of expense, but it is generally carried out by the individual himself for small installations.

It can also have an overpricing of the home insurance; this data depends on the insurance of each, the contract closures, and so on.

These annual fees are re-evaluated every year, resulting in INFLA2 inflation. Since the beginning of self-consumption in France, the TURPE for PV has not increased (**Figure 5**). As the maintenance and insurance tariffs are variable to the situation, it is considered here that the annual investment inflation INFLA2 is 0%.

The total annual investments are summarized in **Table 2**.

The price range for both full resale and self-consumption is shown in **Figure 6**. The blue curves represent the maximum and minimum price to be considered for the annual investment in the total resale and the red curves for self-consumption

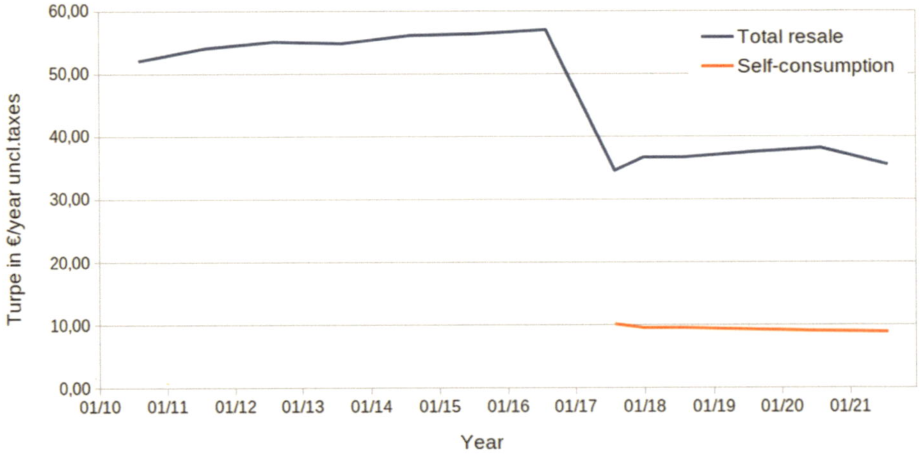

Figure 5.
Evolution of TURPE over the years, for the total resale (blue curves) and self-consumption (red curves).

First Semester 2022	Total resale	Self-consumption with resale	Self-consumption without resale
TURPE	35.45 €/year	8.86 €/year	0€/year
Maintenance	between 0 et 300€/year	between 0 et 300€/year	between 0 et 300€/year
Insurance	between 0 et 100 €/year	between 0 € et 100 €/year	between 0 et 100 €/year
Tax	0 €	0 €	0 €

Table 2.
The different sources of annual investment to be considered for the profitability calculation.

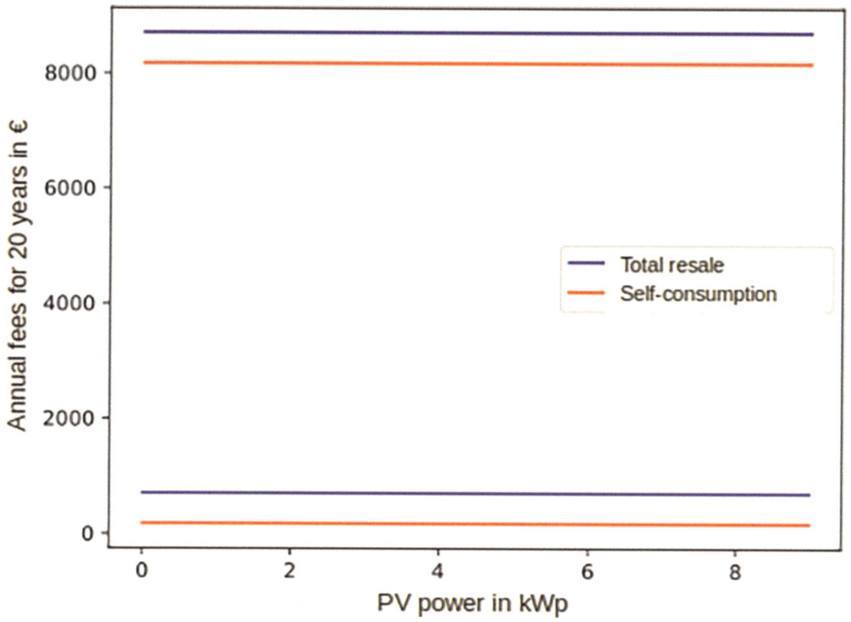

Figure 6.
Extremum for the cumulative annual investment IA over 20 years for total resale (blue curves) and self-consumption (red curves).

according to the possible variations in the TURPE, the price of insurance and maintenance. Taking the extremes, the price of the annual investment is one time cheaper in self-consumption due to the TURPE.

3.3 Solar energy production $Ep_{average}$

The average solar energy production $Ep_{average}$ (kWh/year) follows Eq. (3).

$$Ep_{average}(n, i, o) = \frac{\sum_1^n G_{average}(i, o) \times (\eta(n-1) - \eta(n-1).P) \times S \times PR}{n} \quad (3)$$

Where:
n: 20 years

$G_{average}$(i,o): average global irradiation over a year depending on the tilt i and orientation o of the surface (kW/m^2/year)

η: efficiency of a panel (%)

P: loss of yield of the panels over time (%)

S: total installed area (m^2)

PR: performance ratio (%)

It is necessary to know the average irradiation over a year.

Solar irradiation is a radiometric quantity that measures the amount of solar energy received per unit area after considering atmospheric absorption and scattering. The measured value depends on the tilt of the sensor of the measuring device and its analysis spectrum. The irradiance will allow the calculation of the solar energy produced (kWh). This magnitude will depend, initially, on geographical locations, including the time of sunshine that will be different from one region to another (**Figure** 7), and the latitude, which will play on the path of the sun. For example, in Lille the irradiance for a horizontal surface is on average 1090 kWh/m^2/year, in Toulon 1612 kWh/m^2/year, and in Limoges 1261 kWh/m^2/year. The irradiance data come from the PVGIS software [9], which allows us to know the irradiance according to the geographical area in Europe and North Africa.

It will also vary according to the tilt and orientation of the surface. For example, in Limoges:

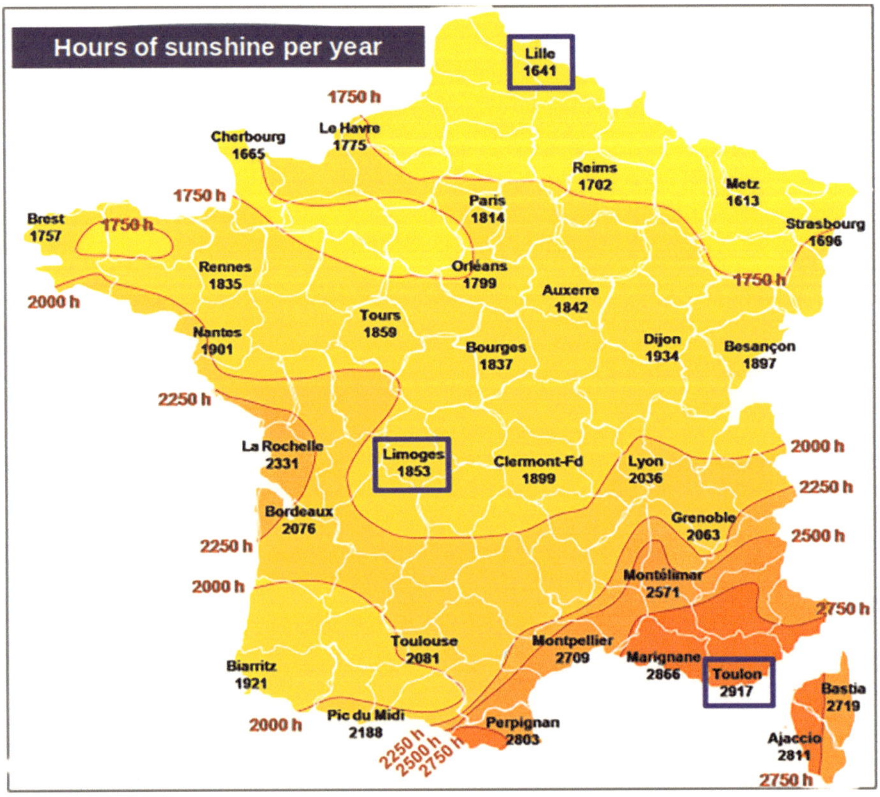

Figure 7.
Average sunshine time in France in the cities of Limoges, Lille and Toulon [8].

- for a tilt of 30° south, an irradiation of 1460 kWh/m^2/year is estimated against 1377 kWh/m^2/year for an irradiation at 60° south.
- for an orientation at 30° east, the irradiation is 1203 kWh/m^2/year.

For Limoges, the most important irradiation is obtained for a tilt of 36°, full south with 1467 kWh/m^2/year on average (**Figure 8**).

Irradiation will also vary over time, daily (day/night, climatic variations) and monthly (seasonality) (**Figures 9** and **10**). Finally, it is never the same from one moment to another, from 1 day to another and from 1 year to another.

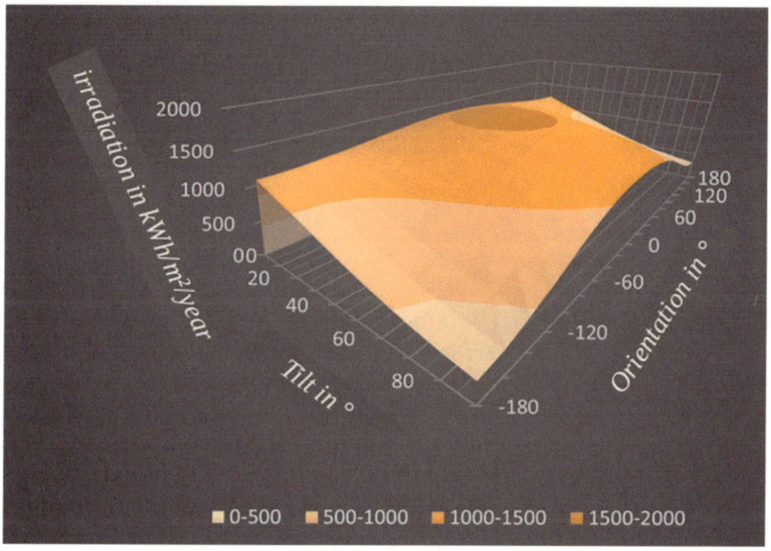

Figure 8.
Variation in irradiation as a function of tilt and orientation according to PVGIS SARAH2 in Limoges in 2020.

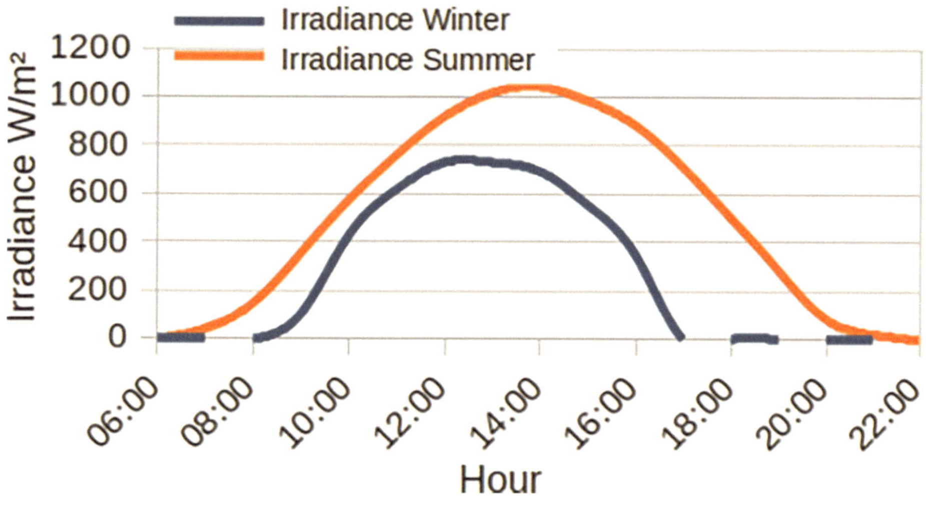

Figure 9.
Daily irradiation variation - 30° south in Limoges.

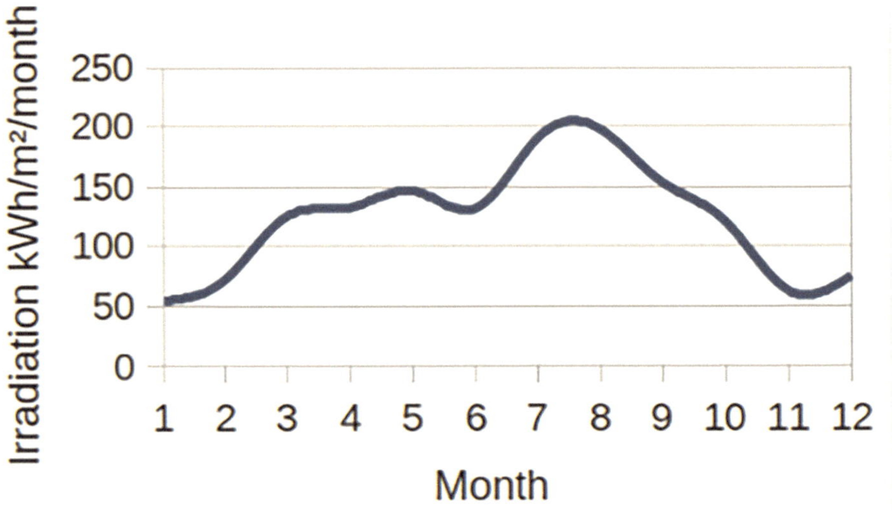

Figure 10.
Monthly variation of irradiation - 30° South in Limoges.

But the calculation of the solar production is not limited to the solar irradiation. It is necessary to know the characteristics of the panels with their yield, their loss over time and the total installed surface. Depending on the technologies and brands, all these parameters vary, ranging from a yield of 19–21% associated with loss of 0.25% per year to 0.55% per year depending on the panels (see **Table 3**).

The PR depends on the losses due to shading, wiring, inverters or microinverters, temperature and so on. According to the study [11], the average PR is 79% whether with microinverters or inverters regardless of the geographical area and the size of the installation. Variations in the PR are possible from 69–91%.

Figure 11 shows the cumulative energy produced over 20 years as a function of the installed peak power. Depending on the panels presented in **Table 3**, the yield and its loss over time and the performance ratio (see **Table 4**), the energy produced will be more or less important. The two blue curves represent the maximum (meaning panels with the best performance ratio and yields) and the minimum (meaning panels with the worst performance ratio and yields) possible energy produced.

3.4 Feed-in tariffs Tvt

For the calculation of the profitability, it is also necessary to know the feed-in price of the Tvt electricity, resold to a supplier. At least until September 2022, the feed-in-tariff, dictated by the CRE (Commission for the Regulation of Energy) in the total resale is 17.89 and 15.21 c€/kWh for installations below 3 and 9 kWp, respectively (**Table 5**). Over time, these feed-in tariffs have been decreasing every 6 months (**Figure 12**). For self-consumption with resale, the feed-in tariff is currently 10 c€/kWh for any installation below 9 kWp and has not changed since the beginning of self-consumption (**Figure 13**).

The feed-in tariff is fixed according to a 20-year locked-in contract with the supplier. It will not change over time for the individual once the contract is signed.

Brand PV	Power Wp	surface m^2	€/(Wp/m^2)	module efficiency %	annual degradation %	Technology
Mono Perc Ecodelta	345	1.69	0.61	20.45	0.5	Monocrystalline Half-cell
Mono Perc DMECG	370	1.82	0.72	20.31	0.55	Monocrystalline Half-cell
Longi LR4 60 HIH	375	1.82	0.72	20.6	0.55	Monocrystalline PERC
Trina Solar Vertex S	400	1.92	0.74	20.8	0.55	Monocrystalline Mutli Busbars
Risen Titan S	400	1.92	0.76	20.8	0.55	Monocrystalline PERC
Shingled Ecodelta	400	1.88	0.77	21.3	0.5	Monocrystalline Shingled
Trina Solar Vertex S	390	1.92	0.78	20.3	0.55	Monocrystalline Mutli Busbars
Trina Honey Mono Perc	375	1.83	0.79	20.5	0.55	Monocrystalline PERC
JNL Solar Full black	320	1.67	0.90	19.78	?	Monocrystalline Full Black
Duonergy Bifacial glass–glass	375	1.85	0.91	20.22	0.4	Monocrystalline Bifacial glass–glass
Bisol Duplex	375	1.86	0.97	20.2	0.55	Monocrystalline Half-cell
Q.Cells Duo G8	362	1.79	1.00	20.1	0.54	Monocrystalline Half-cell without spacing
Mono Perc AE Solar	450	2.71	1.01	20.7	?	Monocrystalline Multi-busbar
Q.Cells Duo G9	335	1.72	1.12	19.4	0.5	Monocrystalline Half-cell without spacing
TrinaSolat Vertex S	400	1.92	1.18	20.8	0.55	Monocrystalline Multi-busbar
TrinaSolat Vertex S Full Black	390	1.92	1.18	20.3	0.55	Monocrystalline Multi-busbar
Q.Cells Duo ML G9	390	1.90	1.23	21.1	0.54	Monocrystalline Half-cell sans without spacing
Hyundai	400	1.96	1.26	20.4	0.55	Monocrystalline Shingled
LG NeON H	375	1.84	1.26	20.4	0.33	Monocrystalline Multi-busbar
Hyundai Ultra Black	400	2.02	1.31	20.4	0.55	Monocrystalline Shingled
SunPoxer Maxeon 5	430	1.90	1.89	22.7	0.25	Monocrystalline Black Contact

Brand PV	Power Wp	surface m^2	€/(Wp/ m^2)	module efficiency %	annual degradation %	Technology
LG NeON R	365	1.73	2.09	21.1	0.6	Monocrystalline Black Contact

Table 3.
The different characteristics of solar panels useful for profitability calculations [10].

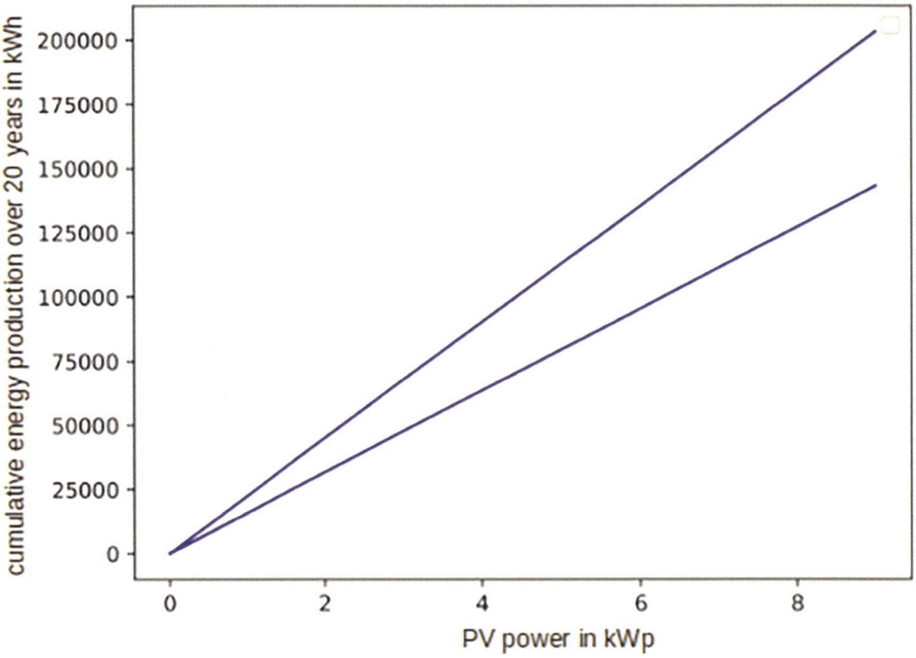

Figure 11.
Extremum for the cumulative solar production over 20 years for an irradiation fixed at 1460 kWh/year (Limoges 30° south).

	Total resale	Self-consumption with resale	Self-consumption without resale
$G_{average}$ (kWh/year)	Between 0 and 2500	Between 0 and 2500	Between 0 and 2500
η (%)	Between 19 and 21	Between 19 and 21	Between 19 and 21
P (%/year)	Between 0.55 and 0.25	Between 0.55 and 0.25	Between 0.55 and 0.25
PR (%)	Between 69 and 91	Between 69 and 91	Between 69 and 91
S (m^2/kWp)	Between 4.4 and 5.2	Between 4.4 and 5.2	Between 4.4 and 5.2

Table 4.
The different sources of variability of the solar production $EP_{average}$.

3.5 Energy tariff TH

For self-consumption, the evolution of the price of electricity will have an impact on the profitability of the installation. In France, to summarize, it is possible to choose

Variable parameters		Total resale	Self-consumption
ID	Installation price (€/kWp)	2770–95 * kWp	2770–95 * kWp
	Connection (€)	600	
	State aid (€/kWp)		380 if ≤3 kWp 280 if ≤9 kWp
IA	TURPE (€/year)	35.45	8.86
	Maintenance (€/year)	0	0
	Insurance (€/year)	0	0
Infla (%)			3
Infla2 (%)		0	0
Tvt (€/kWh)		0.1789 if ≤3 kWp 0.1521 if ≤3 kWp	0.10
THP (€/kWh)			0.1841
$Ep_{average}$ (kWh/year)	$G_{average}$ (kWh/m²/year)	1460 (Limoges 30° south)	1460 (Limoges 30° south)
	η (%)	20.5	20.5
	P (%/year)	0.5	0.5
	S (m²/kWp)	4.9	4.9
	PR (%)	79	79

Table 5.
Data of the variables used for the profitability calculation. The example is for an installation in Limoges with panels arranged 30° south and average characteristics.

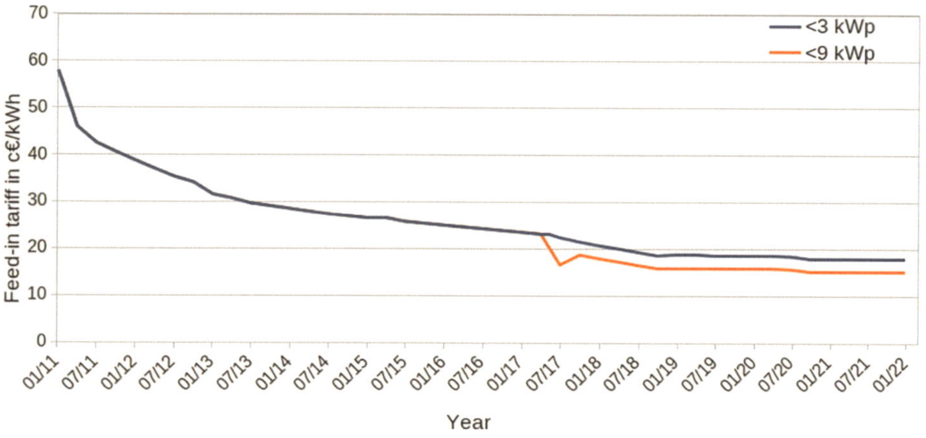

Figure 12.
Evolution of the electricity buy-back (orange curves) and of the state premium (blue curves) over the years for installations below 9 kWp in self-consumption.

an off-peak/peak hour subscription or a basic subscription. The tariffs will be different according to the subscription and according to the time of the day with an off-peak/peak hour subscription.

According to the regulated tariffs of the CRE, the price of electricity has been increasing steadily for both subscriptions as shown in **Figure 14** with a much higher increase for off-peak hours than for peak hours.

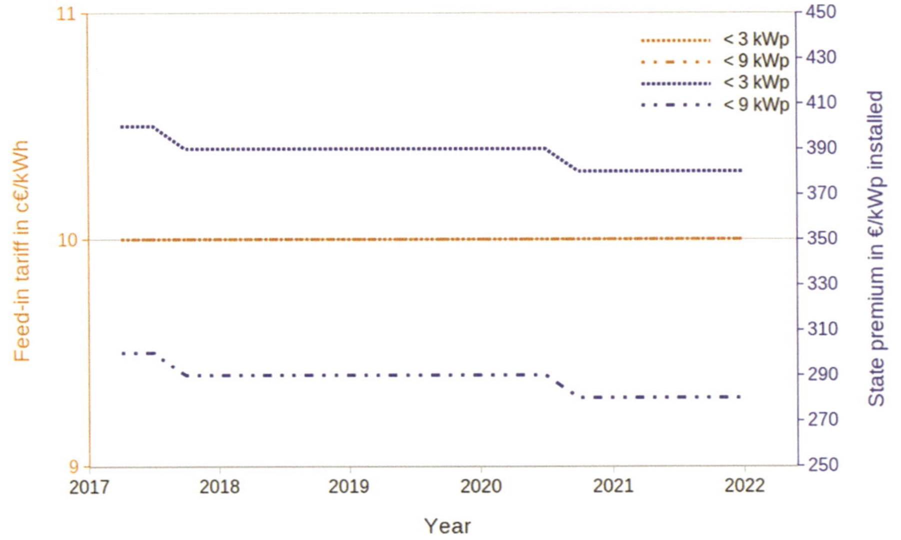

Figure 13.
Evolution of the electricity buy-back over the years for installations below 9 kWp in total resale. Until 2017, the blue and red curves are mixed.

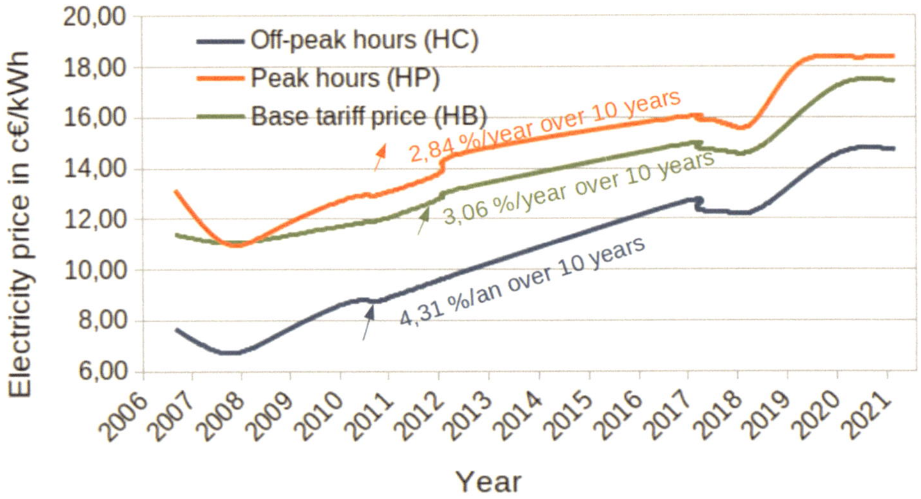

Figure 14.
Evolution of regulated electricity tariffs in France according to the CRE for daytime (peak hours), nighttime (off-peak hours) and basic tariffs (same tariff all day).

In self-consumption, it is preferable to choose off-peak hours HC – peak hours HP contract rather than the basic price because it is during the peak hours HP (most expensive rate) that the solar production will intervene.

The price of electricity is currently 0.1841 €/kWh in HP (between 6 h30 and 21 h30) and 0.1470 in off-peak (between 21 h30 and 6 h30). Please note that the off-peak/peak hours may vary slightly from one area to another.

In addition, for the last 10 years, the price of electricity has been rising steadily, generating an INFLA inflation of about 3%/year for the off-peak/peak hour rates (**Figure 14**).

3.6 Self-consumption rate τ_{self}

The self-consumption rate is the proportion of energy directly consumed on site. This parameter is only useful for calculating the profitability in self-consumption.

The annual self-consumption rate τ_{self} follows Eq. (4).

$$\tau_{self}(i,o) = \frac{Productionconsumedonsite}{Totalproduction} = 1 - \frac{Surplus(i,o)}{Totalproduction(i,o)} \tag{4}$$

Where

$$Totalproduction(i,o) = Epaverage(i,o) \tag{5}$$

$$Surplus(i,o) = \sum\nolimits_{1year}(Ep(i,o) - consumption) \tag{6}$$

The surplus being the moment when the production is higher than the consumption (Eq. 6).

For the calculation of surplus, it is necessary to know the consumption and production or by substitution the irradiation (Eq. 5), at an identical time scale and the shortest possible time to have a rate of self-consumption as realistic and accurate as possible.

To try to take at best the consumption variabilities, the Linky communicating meter will allow to recover the consumption data and thus the load curves by semi-hourly steps. For the example of the profitability calculation of this article, the load curve recovered is for an all-electric household (water heater, electric heating) of four people and 100 m^2, in the countryside near Limoges, for the year 2019 with an average consumption of 15,000 kWh/year. The load curves are in red in **Figures 15–17**. Note that, for all other houses, the load curve will be completely different because it depends on the household appliances, the heating system, the insulation, the surface, the number of people, the habits of each person and so on.

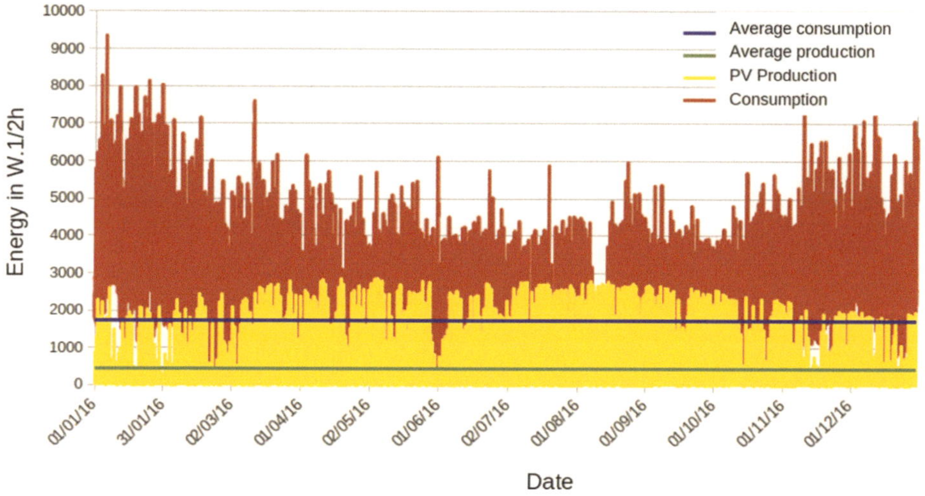

Figure 15.
Solar production for a 3 kWp installation at 30° south in 2019 in Limoges (yellow curve) and load curve in 2019 for a house of four people, all electric, in Limoges (red curve). The green curve represents the annual average of production over the year 2019.

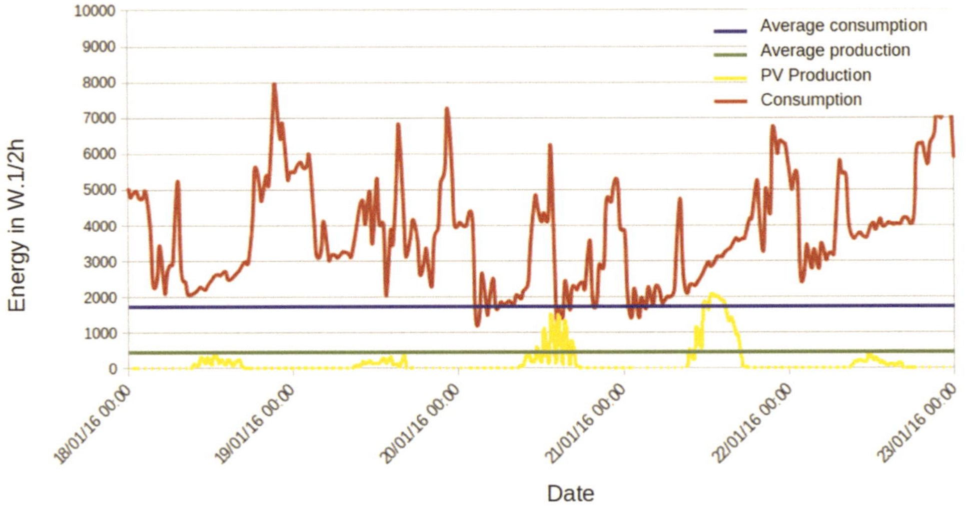

Figure 16.
Zoom-in January solar energy for 2.84 kWp 30° South in Limoges in 2019 and load curve 2019 for a house of four persons, all electric, in Limoges.

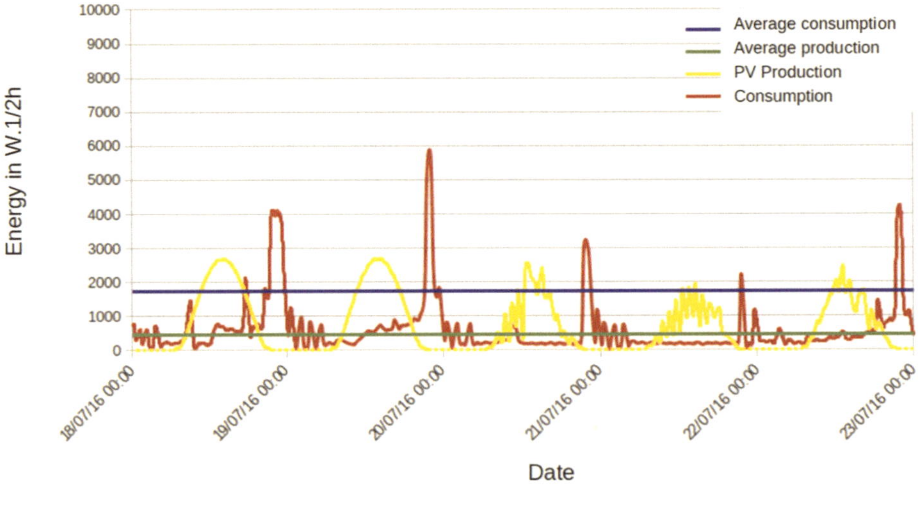

Figure 17.
Zoom-in July solar energy for 2.84 kWp 30° south in Limoges in 2019 and load curve 2019 for a house of four persons, all electric, in Limoges.

For the production variability, PVGIS [9] will give the irradiance for any location, tilt and orientation in Europe and North Africa following the calculation methods detailed in [12–15]. The SARAH [16] and ERA5 [17] data give irradiances in hourly steps from 2005 to 2020. To reduce the hourly step to a semi-hourly step, the values of SARAH (satellite passage at 00:00–1:00 - …) and ERA5 (satellite passage at 00:30–1:30 - …) have been switched for the year 2019. The production curve for 3 kWp 30° south in Limoges is represented by the curves in yellow in **Figures 15–17**.

Moreover, the rate of self-consumption will vary according to the layout of the installation. Indeed, the surplus will be less important if the production is less and better adapted to the consumption curve. For example, the self-consumption rate is more important if the panels are oriented east rather than south (**Figure 18**) or if the installation is in a less sunny geographical area (**Figure 19**). Be careful that a higher self-consumption rate does not necessarily mean a higher yield (**Figure 20**).

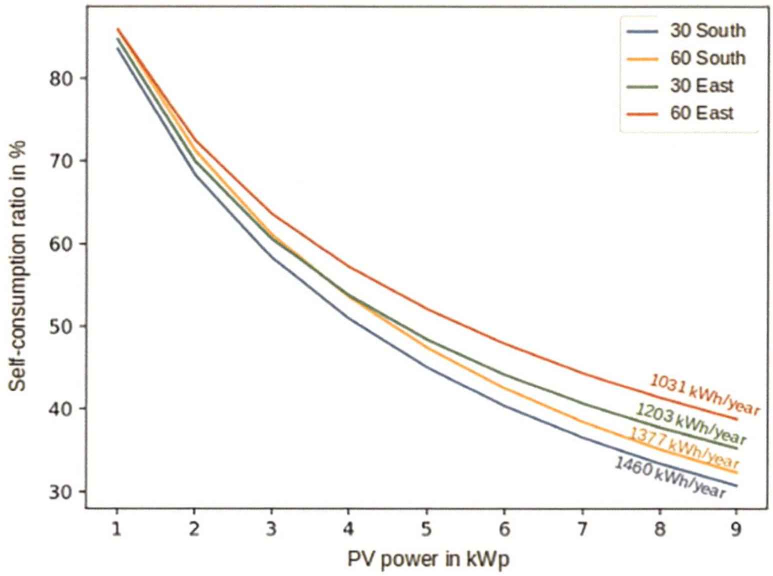

Figure 18.
Self-consumption rate according to the installed power according to different tilts (30° and 60°) and orientations (south or east) for the city of Limoges and for a typical consumption of a house of 100 m² for 4 persons with electric heating.

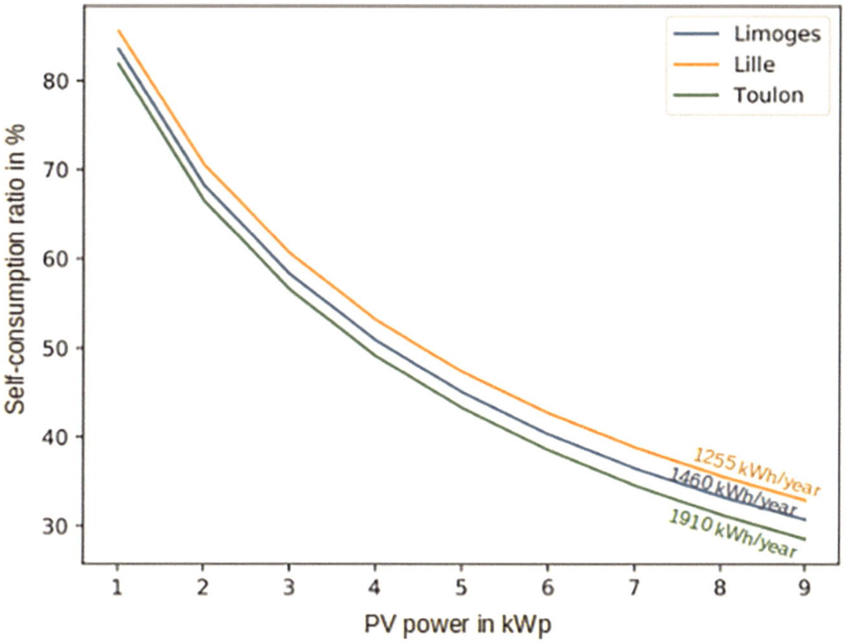

Figure 19.
Self-consumption rate according to the installed power for different cities (Limoges, Lille and Toulon) for installations inclined at 30° and oriented south and for a typical consumption of a house of 100 m² of 4 persons with electric heating.

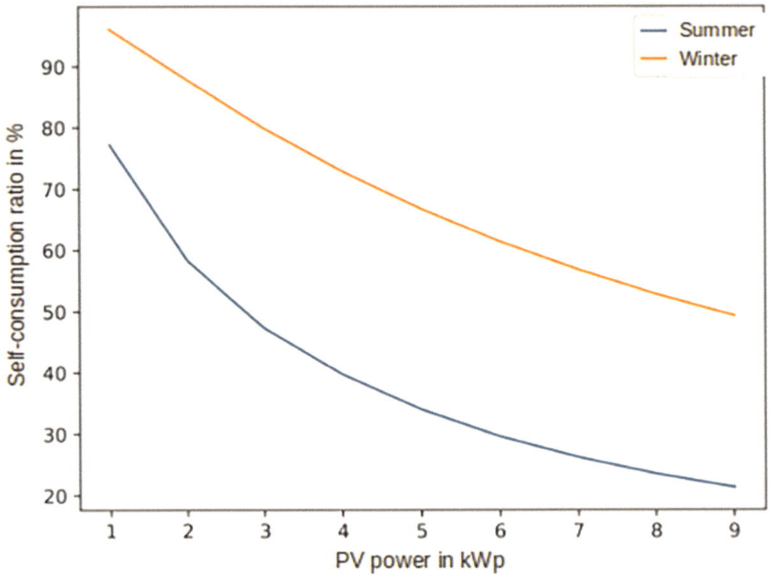

Figure 20.
Self-consumption rate according to the installed power separated in two seasons, summer (from April to September) and winter (from October to March), for the city of Limoges and for a typical consumption of a house of 100 m^2 for four persons with electric heating.

4. Results

Starting from Eqs. (1) and (2) for total resale and self-consumption, respectively, as well as Eq. (3) for the average energy produced, the values of each variable are shown in **Table 5** and will be used for an example of profitability curve.

The values of the variables are chosen for a specific case, but they are adjustable according to the many parameters exposed in part III. The example here is for an installation in Limoges with panels placed 30° south and average characteristics of what is currently found. For self-consumption, the load curve is recovered for an all-electric household (water heater, heating) of four people and 100 m^2, in the country-side near Limoges, for the year 2019 consuming an average of 15,000 kWh/year.

For any installed power from 1 to 9 kWp under the conditions of an irradiation of 1460 kWh/m^2/year, self-consumption is more profitable, ranging from a gain over 20 years of 2500€ for 1 kWp to 14,500€ for 9 kWp. On the other hand, in the total resale of the installations, lower than 1 kWp is not profitable but can still generate gains of 11,500€ for 9 kWp (**Figure 21**). Consequently, the year from which the installation is amortized arrives well before (**Figure 22**) for self-consumption, approximately 12 years for any installed power against 15–16 years for the total resale on small installation and being able to go down to 12 years for a 9 kWp. On the other hand, the range of profitability for the same irradiation is large according to the variables. The curves of the extremes were obtained by starting from **Figures 4**, **6** and **11**. In the most pessimistic case, after 20 years of the contract, the installation will still not be amortized whatever the installed power and the typology of installation.

In addition, to compare the gain from solar production to a $P_{financial}$ investment over several years, Eq. (7) is used.

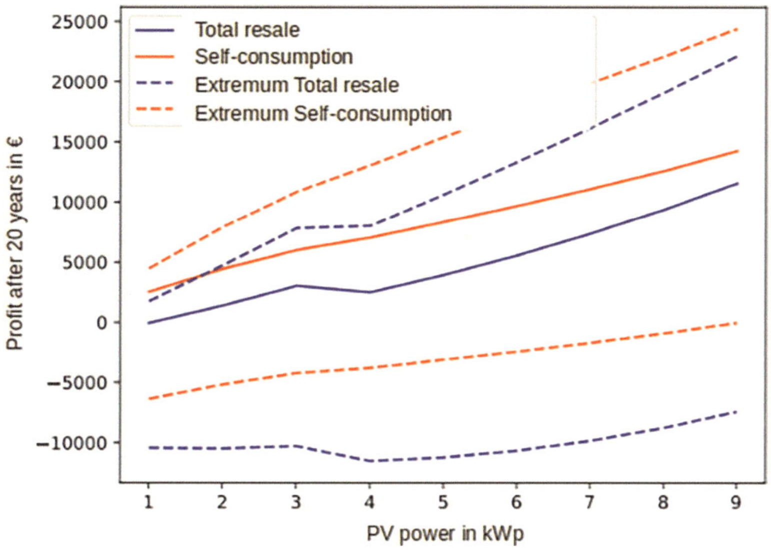

Figure 21.
Comparison of the gains reported by an installation after 20 years of the contract for different installed powers for total resale and self-consumption in Limoges disposed 30° south according to the average variables and their extremums.

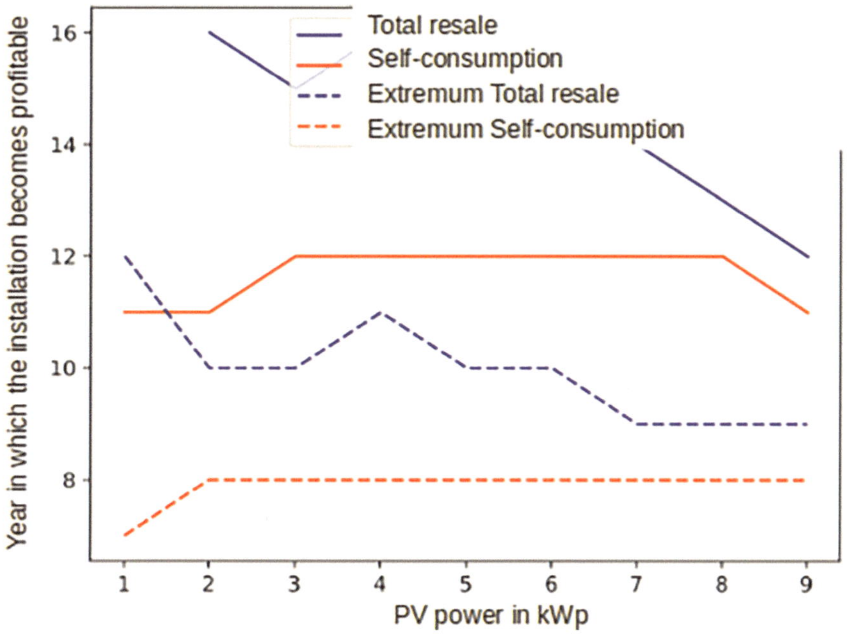

Figure 22.
Comparison of the years from which the photovoltaic installation becomes beneficial for different installed powers for the total resale and self-consumption in Limoges disposed 30° south according to the average variables and their extremums.

$$P_{financial} = \left[\left(\frac{Gain(n,i,o) + ID}{ID} \right) || \frac{1}{n} - 1 \right] \times 100 \quad (7)$$

The financial investment $P_{financial}$ shows the interest rate that the initial investment would yield if it had been put in the form of an investment in percentage per year during the 20 years of the contract.

Contrary to the gain, for self-consumption, a small installation corresponds to a more interesting financial investment that can go in our example to an equivalent investment of 3% per year for 20 years for the individual (red curve in **Figure 23**). On the other hand, for total resale (blue curve in **Figure 23**), the larger the installation, the more interesting the investment becomes over 20 years.

Figures 24–26 show, respectively, the 3 criteria of profitability (earnings, year of amortization and interest rate) for installations of 3 and 9 kWp in the total resale by starting again from the criteria of **Table 5**.

These curves allow us to show that the profitability in the total resale is more important for big installations and for higher irradiations, depending as well on the geographical zone or on the orientation and tilt of the installation. Examples of profitability have been placed according to geographical areas well exposed (Toulon and Lille) and according to orientations (east and south) and tilts (30° and 60°) of the different panels.

For self-consumption, **Figures 27–29** show the profitability according to the same three criteria (earnings, year of amortization and interest rate) as before for a 9 kWp photovoltaic installation. The higher the average irradiation over a year and the higher the self-consumption rate, the higher the profitability.

In the figures, different points have been placed corresponding to different geographical areas (Limoges, Lille or Toulon), orientations (south or east) and tilts (30° or 60°) of the panels and the consumption of the house according to an electric heating or not by making the hypothesis that the rate of self-consumption of a house

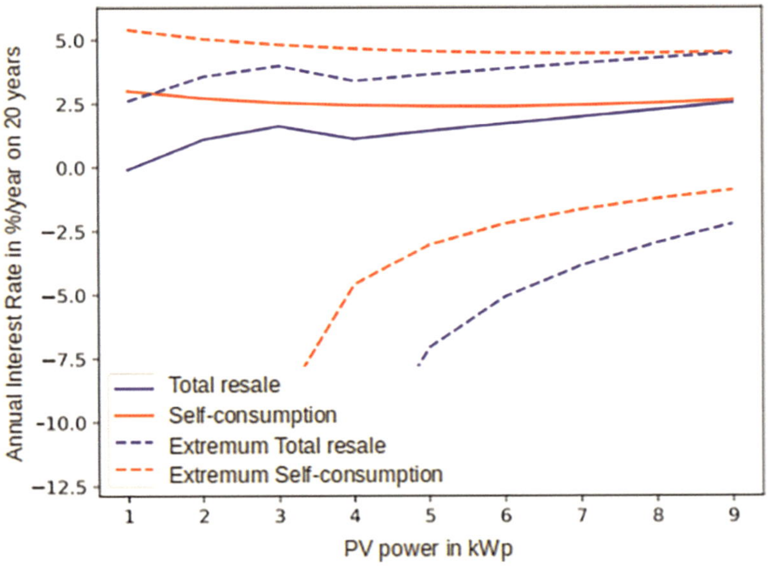

Figure 23.
Comparison of the interest rates that an installation would earn after 20 years of the contract for different installed powers for the total resale and self-consumption in Limoges located at 30° south.

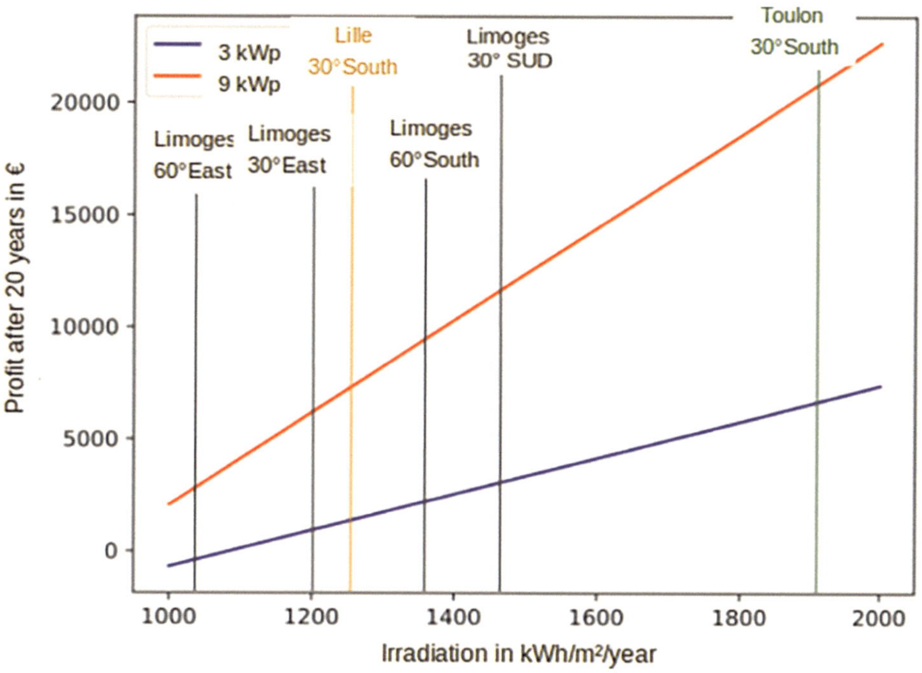

Figure 24.
Comparison of the gains reported after 20 years of the contract for installations of 3 and 9 kWp according to different irradiances corresponding to installations with different orientations, tilts, and geographical areas for total resale.

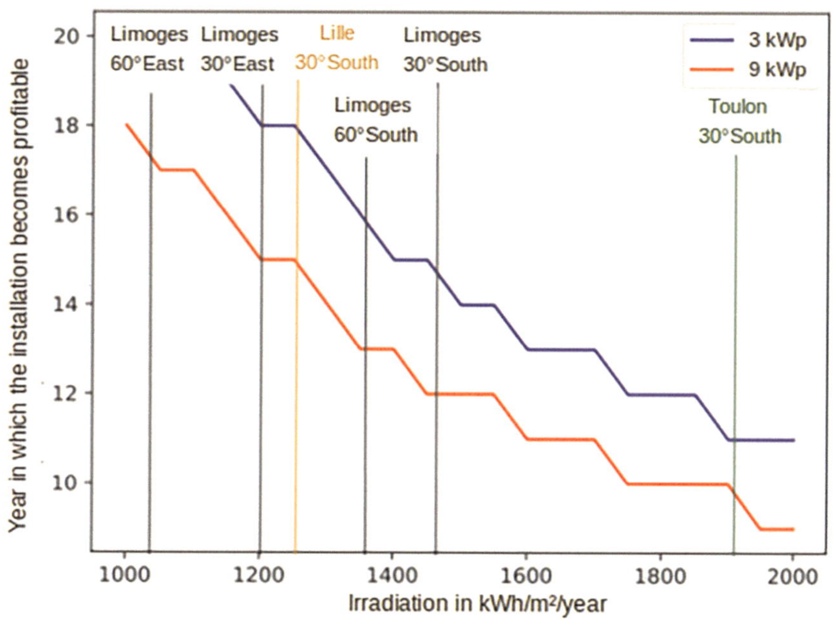

Figure 25.
Comparison of the years from which the photovoltaic installation becomes beneficial for installations of 3 and 9 kWp according to different irradiances corresponding to installations with different orientations, tilts and geographical areas for the total resale.

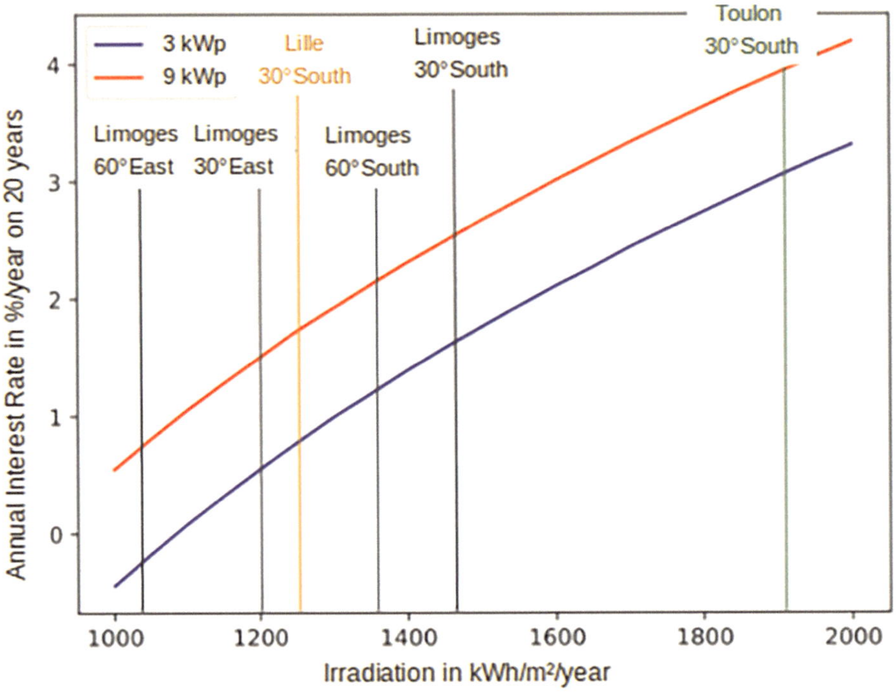

Figure 26.
Comparison of the interest rates that an installation would earn after 20 years of the contract for installations of 3 and 9 kWp according to different irradiances corresponding to installations with different orientations, tilts and geographical areas for the total resale.

without electric heating was the same in the year as the rate of self-consumption of the summer part (**Figure 20**).

For example, an installation in self-consumption of 9 kWp, located in Limoges (center of France), inclined 30° and oriented south with a consumption of 15,000 kWh/year (a house with electric heating) corresponds to an average irradiation of 1460 *kWh/year and a rate of self-consumption of 30.8%. In this situation, the gain after 20 years* will be 14,250 €, allowing to make the installation profitable after 11 years and corresponding to an investment of about 2.63% per year over 20 years. For the same installation but with a consumption not using electric heating, the irradiation is the same, but the rate of self-consumption becomes weaker, of 21.3%. Thus, the profitability will be less because the gain will be 11,370 €, allowing us to make the installation profitable after 12 years and corresponding to an investment of about 2.09% per year over 20 years.

Also, the fact that the orientation and the tilt of the panels differ from the optimum (30° south) increases the rate of self-consumption but decreases the irradiation, not allowing us to have a more important profitability, quite the contrary.

5. Conclusions

Self-consumption is developing more and more in the past years instead of the installations in the total resale. And for good reason, in addition to participating in the

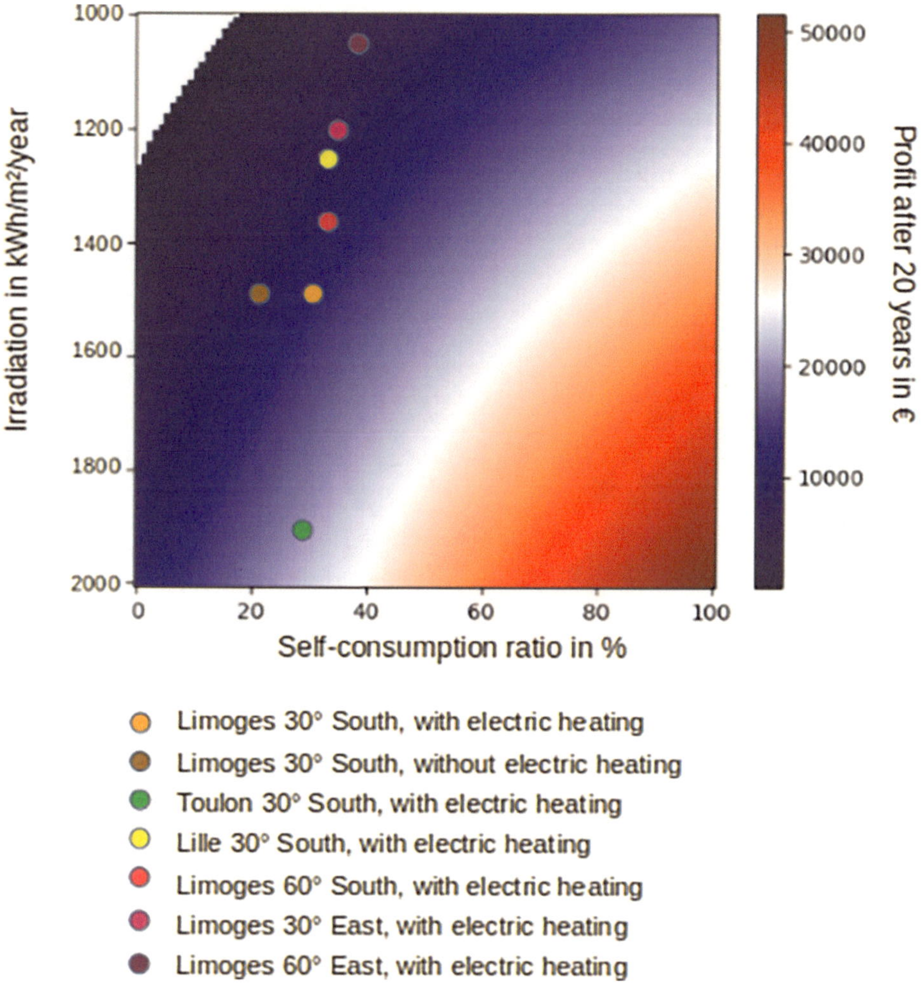

Figure 27.
Comparison of the years from which the 9 kWp PV installation in self-consumption becomes beneficial according to different irradiances and self-consumption rates and based on the variables presented in ***Table** 5*.

energy transition, a self-consumption installation is an interesting alternative for a monetary investment.

Indeed, currently with a buy-back price of electricity that is increasingly low for the total resale, self-consumption is in most cases more profitable for installations at home ranging from 1 to 9 kWp.

The benefits are more advantageous in the case where the irradiation is maximum on average over the year, 30 to 40° south, although the correlation consumption/production is the worst.

Moreover, the calculations of profitability were made for 20 years because it is the time of the contract of repurchase of the electricity proposed by EDF. But at the end of this contract, nothing prohibits to renew it. The initial investment being already amortized and having no repurchase of material (life span of the panels higher than 30 years and that of the micro-inverters of 25 years), the profitability will be only better than that calculated here.

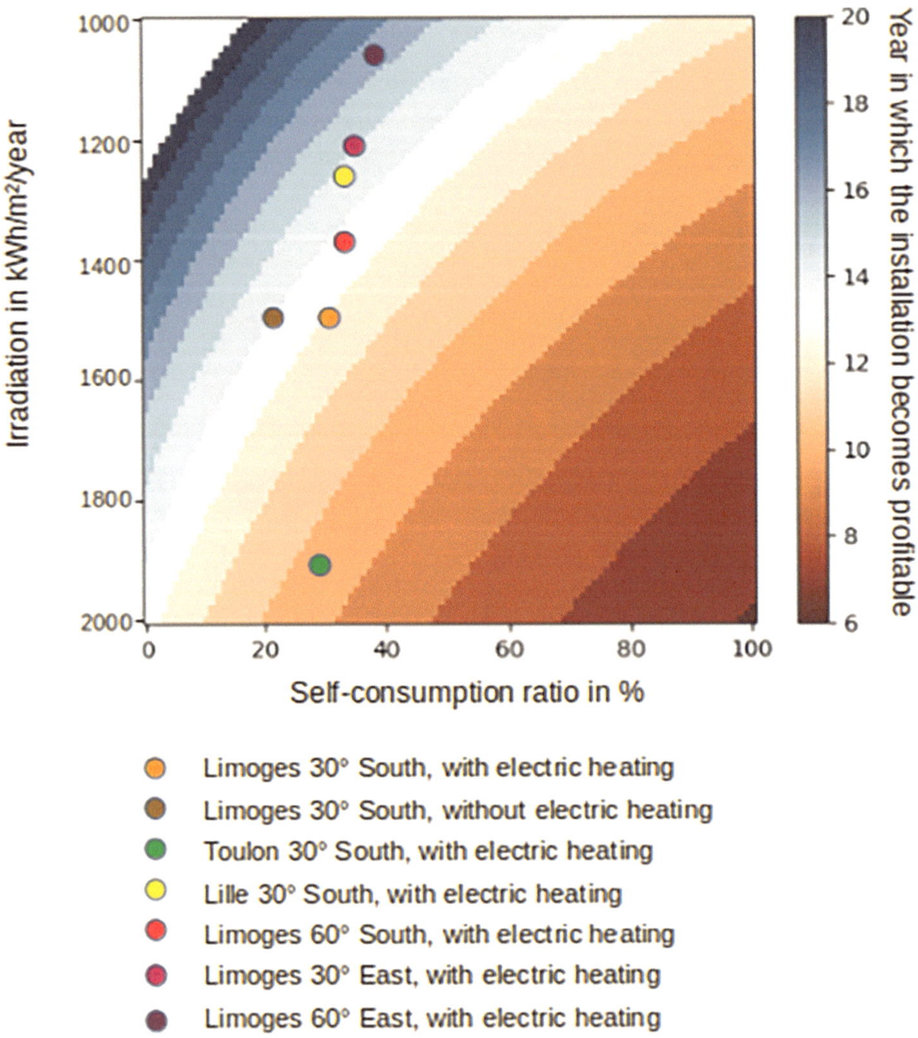

Figure 28.
Comparison of the gains reported after the 20 years of the contract for a 9 kWp installation installed in self-consumption according to different irradiances and self-consumption rates and using the variables presented in ***Table** 5*.

Also, the profitability in self-consumption can be improved by using an energy management. Indeed, if appliances are used at the most opportune moments, when there is a surplus, the rate of self-consumption, the quantity of energy produced that is directly consumed on site and not lost or resold at low cost, will only be better. If for example, the water heater, the washing machine and so on are launched at the time of the surplus to the link of evening, the rate of self-consumption will increase and consequently the profitability will improve.

Note that for each calculation of profitability in self-consumption, the calculations are entirely to be redone because each household has its own consumption habit.

Thus, it would be interesting to make profitability calculations for different load curves: household without electric heating (represents 80% of the houses), household of two persons, household of six persons, household of retired person and so on. The electrical energy can be divided or multiplied by 2 or 3, with different consumption habits.

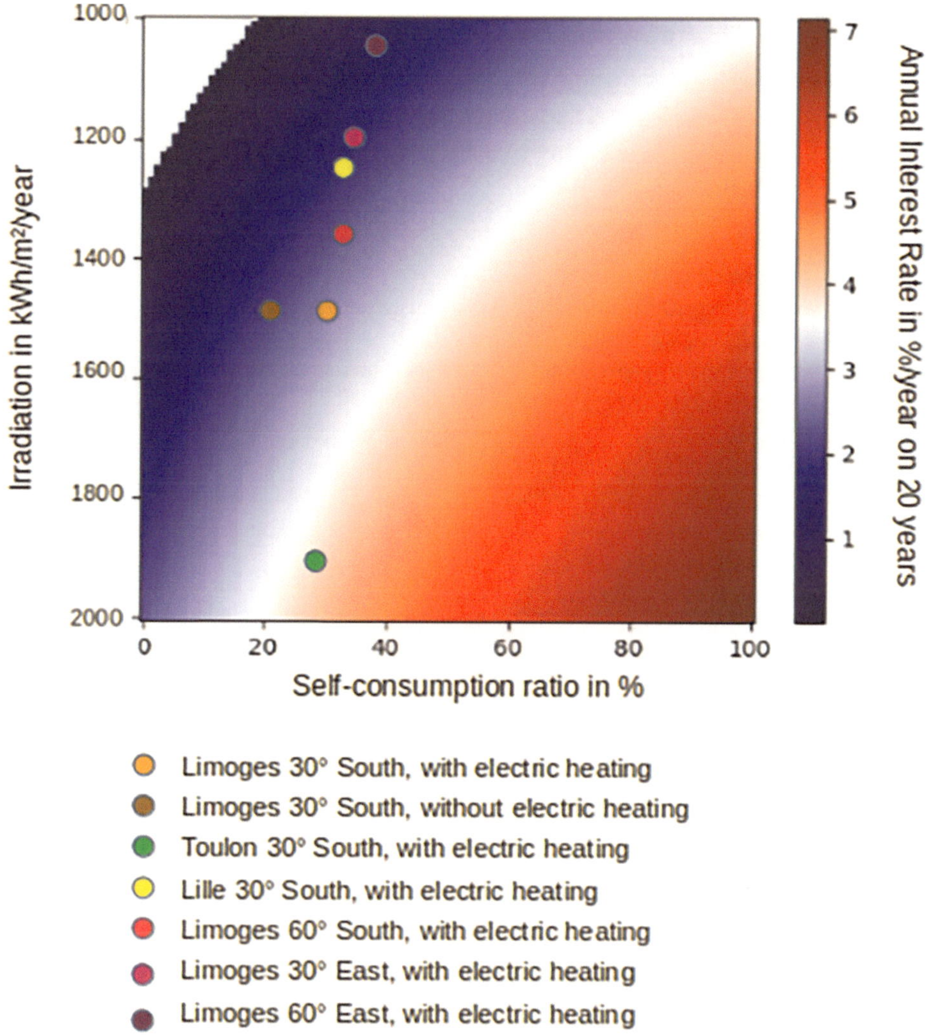

Figure 29.
Comparison of the interest rates that an installation would earn after 20 years of the contract for a 9 kWp installation installed in self-consumption under different irradiances and self-consumption rates and based on the variables presented in ***Table** 5.*

Finally, investing in photovoltaic seems to be interesting in all cases (total resale or self-consumption with sale of surplus). But to privilege the self-consumption seems to be the most judicious especially as the inflation of electricity has great chance to increase more quickly than in the past, having as a direct effect an increase in the profitability of an installation in self-consumption.

Acknowledgements

This work is part of the UNIVEERS and LIMBATT project, co-financed by the European Union in the framework of the FEDER-FSE 2014-2020 and by the Nouvelle-Aquitaine region.

Conflict of interest

"The authors declare no conflict of interest.

Appendices and nomenclature

τ_{self}	self-consumption rate
CRE	energy regulation commission (Commission de régulation de l'énergie in French)
$Ep_{avearge}$	average energy produced over a year based on an average irradiance
$G_{average}$	average irradiance
THP	peak hour tariff
Tvt	feed-in tariff
ID	initial investment
IA	annual investment
Infla	inflation of electricity
Infla2	inflation of the annual investment price
TURPE	tariffs for the use of public electricity networks
kW	kilowatt
kWh	kilowatt hour
kWp	kilowatt peak
PR	performance ratio
PV	photovoltaic panel
W	watt

Author details

Quentin Lagarde[1*], Bruno Beillard[1], Serge Mazen[1] and Julien Leylavergne[2]

1 Research Institute XLIM, University of Limoges, Limoges, France

2 University Institute of Technology, University of Limoges, Limoges, France

*Address all correspondence to: quentin.lagarde@unilim.fr

References

[1] OpinionWay pour ADEME. Attitude des Français á l'égard de la qualité de l'air et de l'énergie, 8ème vague de l'enquête annuelle sur la qualité de l'air, les énergies renouvelables et les économies d'énergie dans le logement. Dec 2021. 47 p. Available from: www.ademe.fr/mediatheque

[2] Statinfo - solaire photovoltaïque_2019T4. Available from: https://www.statistiques.developpement-durable.gouv.fr/publicationweb/263 [consulté le juill. 01, 2020]

[3] ObservER-Barometre-EnR-Electrique-France-2019.pdf. Accessed: Jul 01, 2020. [Online]. Available from: http://www.energies-renouvelables.org/observ-er/html/energie_renouvelable_france/ObservER-Barometre-EnR-Electrique-France-2019.pdf

[4] Radhwane N, Oussama RM, Salem C, et al. Analyse et Simulation de Performance d'un Central Photovoltaïque Raccordé au Réseau. [thèse de doctorat]. Universite Ahmed Draia-Adrar. 2021

[5] Rasmus L, Widén J, Daniel N, et al. Photovoltaic self-consumption in buildings: A review. Applied Energy. 2015;**142**:80-94

[6] Barry F, Margolis R, Seel J. Comparing photovoltaic (PV) costs and deployment drivers in the Japanese and US residential and commercial markets. National Renewable Energy Lab. (NREL), Golden, CO (United States). 2016

[7] Jean-Francois C, Christine C, Edwige C, et al. Deliberation Nr 2020-318: Deliberation of the Commission for the regulation of energy of the 17 December 2020 bearing project of decision of the price of use of public networks of electric power distribution (Turpe 6 HTA-BT). Deliberation Nr 2020-314: Deliberation of Commission for the regulation of energy of the 17 December 2020 bearing project of decision on the price of use of public networks of electric power transport (Turpe 6 HTB). 2020

[8] L'ensoleillement en France. Available from: https://www.lepanneausolaire.net/l-ensoleillement-france.php [accessed July 02, 2020]

[9] JRC Photovoltaic Geographical Information System (PVGIS) - European Commission. Available from: https://re.jrc.ec.europa.eu/pvg_tools/en/tools.html#MR [accessed June 30, 2020]

[10] Photovoltaïque, panneaux solaires et kits autonomes - Wattuneed. Available from: https://www.wattuneed.com/?utm_source=criteo [accessed July 02, 2020]

[11] Quentin L, Bruno B, Serge M, et al. Performance ratio of photovoltaic installations in France: Comparison between inverters and micro-inverters. Journal of King Saud University-Engineering Sciences. 2021

[12] Huld T, Müller R, Gambardella A. A new solar radiation database for estimating PV performance in Europe and Africa. Solar Energy. 2012;**86**(6): 1803-1815

[13] Amillo AG, Huld T, Müller R. A new database of global and direct solar radiation using the eastern meteosat satellite, models and validation. Remote Sensing. 2014;**6**(9):8165-8189

[14] Urraca R, Huld T, Lindfors AV, Riihelä A, Martinez-de-Pison FJ,

Sanz-Garcia A. Quantifying the amplified bias of PV system simulations due to uncertainties in solar radiation estimates. Solar Energy. 2018;**176**: 663-677

[15] Yordanov GH. Relative efficiency revealed: Equations for k 1–k 6 of the PVGIS model. In: 2014 IEEE 40th Photovoltaic Specialist Conference (PVSC). 2014. pp. 1393-1398

[16] Müller R, Pfeifroth U, Träger-Chatterjee C, Trentmann J, Cremer R. Digging the METEOSAT treasure—3 decades of solar surface radiation. Remote Sensing. 2015;7(6):8067-8101

[17] Urraca R, Huld T, Gracia-Amillo A, Martinez-de-Pison FJ, Kaspar F, Sanz-Garcia A. Evaluation of global horizontal irradiance estimates from ERA5 and COSMO-REA6 reanalyses using ground and satellite-based data. Solar Energy. 2018;**164**:339-354

Chapter 7

Utilization of MOSFET Transistor to Characterize PV Panels under Dust: Study Area Agadir-Morocco

Abdellah Asbayou, Lahoussine Bouhouch, Ismail Isknan and Ahmed Ihlal

Abstract

The accumulation of dust on the surface of photovoltaic (PV) modules reduces the intensity of the light transmitted through the cover glass, and therefore, the amount of energy generated by the solar cells. This issue, known as soiling. Affects PV systems worldwide, causing power losses as high as 70% in the worst scenarios This chapter presents an electro-optical investigation of the dust accumulated on the PV panel in the study area of Agadir-Morocco, by using a MOSFET transistor as load to truck the IV and PV characteristics of SX330J. For this purpose, Experiments of soiling effects on the performances of a PV panel have been performed using dust collected from two sites in the region of Agadir, Morocco: Adrar (AD) and Halieutic-Parc (HP). The results suggest that measuring the optical transmittance of the soiling accumulated on a PV glass can give enough information to quantify the impact of soiling on the energy production.

Keywords: solar panel, characterization, dust effect, transmittance, IV characteristics

1. Introduction

The production of electricity from solar PV has become an increasingly crucial source of energy [1], conversion of solar energy into electricity is required by means of solar PV cells. These photovoltaic cells are usually made from silicon, which remains the most technologically and industrially advanced field [2]. Among the environmental factors that can have a detrimental impact on the performance of PV cells is dust accumulation.

Dust accumulation on PV modules affects solar PV plants worldwide. In fact, the dirt deposited on the surface of the PV panels consists of mineral dust, aerosols, pollen, fungi and/or other contaminants [3]. The dust particles absorb, disperse, and reflect a proportion of the incident sunlight, thus reducing the intensity of light reaching the active part of the PV cell. Therefore, in some regions, a power degradation of more than 50% has been reported in the literature [3, 4].

Determining the physical properties of dust (e.g., size, geometry, weight, and type of pollutant) provides information's on the degradation of PV module performances

IntechOpen

[5]. It is important to understand the relationship between the dust amount and its impact on the dispersion and transmission of sunlight. Several authors found that the accumulated dust on solar panel are dominated by the region [6]. In the literature, the authors [7] discern that the presence of dust on the surface of PV modules reduces their power by half if they are exposed for 6 months without cleaning. The dust accumulation process is directly related to the wind movement; depending on its strength and speed, it gradually covers the entire surface of the PV panel with several thin layers of dust [7, 8]. On the other hand, in southern Spain, a study of [8] reported that the daily energy losses due to dust were about 5% for 12 months, but in Cyprus, the authors [9] estimated a 13% power output decrease due to dust recorded during 12 months of data recording.

The degradation of the efficiency of solar panels is mainly due to the decrease in optical transmittance due to the accumulation of dust on the upper part of the PV panel's [10]. The dust accumulation process is directly related to the wind movement; depending on its strength and speed, it gradually covers the entire surface of the PV panel with several thin layers of dust [7, 8, 11].

In this chapter, an electro-optical investigation of the dust accumulated on the PV panel in the study area of Agadir-Morocco has been presented, by using a MOSFET transistor as load to truck the IV and PV characteristics of SX330J. The experimental results was compared to the simulation, and good agreement was founded.

This chapter is organized as follows: After introduction, in Section 2; we present the study area. Then in Section 3; we give the methodology used in this investigation. Then, we present the electrical modeling and simulation of PV panel under dust in Section 4, the results and discussions in the Section 5. Finally, we conclude our chapter with a conclusion on our investigations, while proposing some perspectives, to further develop this work.

2. Experimental area

The region of Agadir, the district capital of Souss-Massa, is located on the southern Atlantic side of Morocco in the foothills of the Anti-Atlas. The area chosen for our investigations, is situated at the following geographical coordinates: 30°25′North, and 9°36′West. The type of dust targeted were collected from two sites in the plain of Souss-Massa, namely: the site of Adrar AD (30°25′37.6″ N, 9°32′24.1″ W) and the site of Halieutic-Park HP (30°24′48.4″ N, 9°24′04.5″ W) (**Figure 1**).

The meteorological data were acquired from a station located in Agadir, University IBN ZOHR, Department of Geology, for a period of 6 months from 24/02/2021 to 06/09/2021. The representative data of irradiance, and ambient temperature are shown in **Figure 2**.

Besides the interesting solar potential ranging from 0.8 to 1.2 kWh/m^2, the study area is hosting intensive industrial, touristic, and agricultural activities. Such activities are likely to contribute to the soiling of PV panels. In **Figure 2**, the average of temperature (green curves) in Agadir is respectively between 10°C and 45°C.

3. Methodology

In order to study the electrical characteristics (I–V) of PV a panels, depending on the amount of the soiling accumulated on a PV glass, two soil samples were collected

Figure 1.
Moroccan solar energy field and experimental study sites.

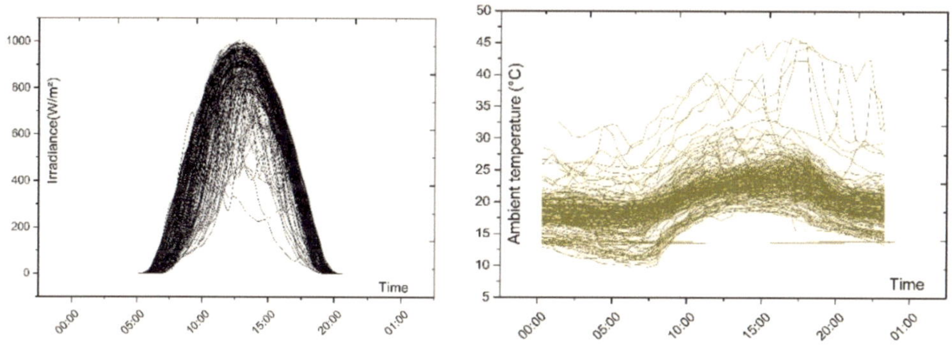

Figure 2.
Meteorological data of Agadir area between 24/02/2021 and 06/09/2021.

from the area of Agadir-Morocco in the Souss-Massa plain. The following experimental protocol was adopted. Upon collected, the dust samples were sieved to get a fine powder of less than 150 μm particles diameter. The particles were then placed on the solar module SX330J as a test panel. **Table 1** shows the electrical parameters of the PV module supplied by the manufacturer.

Two analytical techniques were used to reach the electro-optic properties of the collected dust. JASCO V730 UV-Vis spectrophotometer was used to compare the evolution of optical transmittance of light (300–1100 nm) as a function of dust density. The electrical circuit proposed in this article is realized with an Arduino UNO board, as a unit of acquisition, control, and transfer of data (Ipv and Vpv), in real time, to a computer via the PLX-DAQ tool [12, 13]. The Arduino module generates a PWM signal, then this signal is injected to an RC filter to control the variation of the voltage VGS of the MOSFET [14].

Parameters	Value
Pmax: Experimental maximum power (W)	30 W
Vmp: Maximum voltage (V)	16.8 V
Imp: Maximum current (A)	1.78 A
Isc: Short-circuit current (A)	1.94 A
Voc: Open-circuit voltage (V)	21.0 V
α: Temperature coefficient	(0.065 ± 0.015) %/°C
β: Temperature coefficient	$-(80 \pm 10)$ mV/°C
Temperature coefficient of power	$-(0.5 \pm 0.05)$ %/°C

Table 1.
PV panel SX330 parameters at STC (1000 W/m² and 25°C).

The measurements of the current (Ipv) and the voltage (Vpv) at the output of the PV panel under test are performed using Arduino compatible current and voltage sensors. For the temperature of PV panel, it is measured by a thermocouple, while the irradiance is measured by the FI 109SM solarimeter (**Figure 3**).

The assumptions of the work focus on:

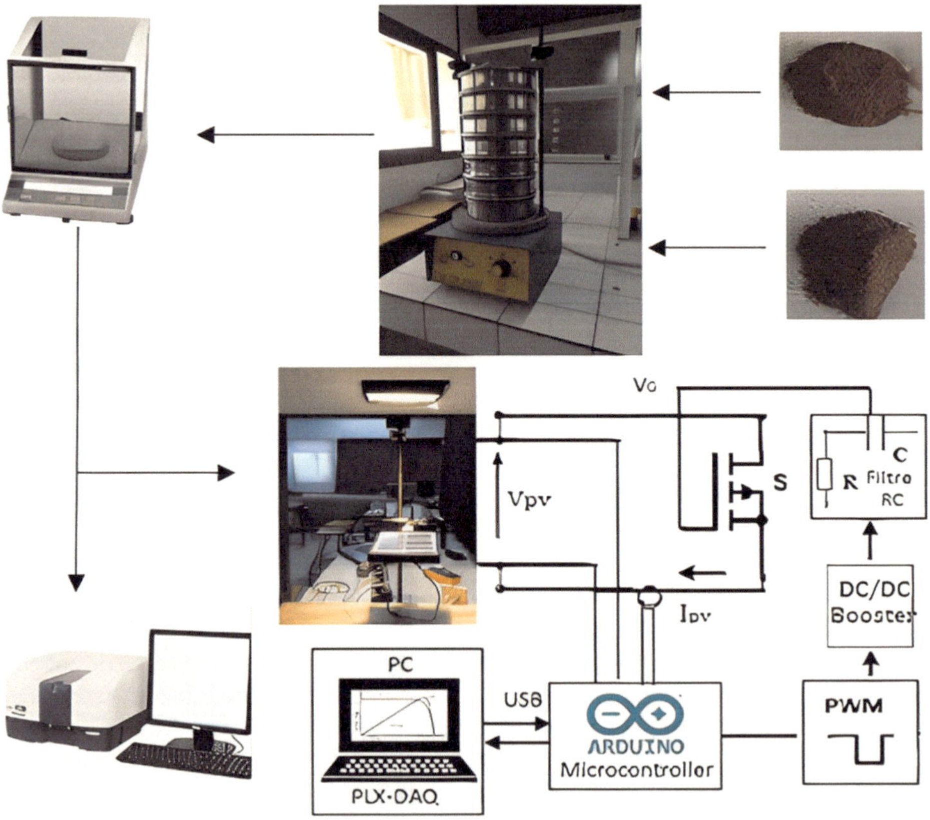

Figure 3.
Realized test bench of PV panel under dust effect.

1. Normal incidence of light is on the solar panel.

2. The size of the dust considered in the study is less than 150 μm.

3. The deposition of the dust is done mechanically in order to have a uniform distribution on the PV panel.

For the accuracy of the material used, the JASCOV-730 Spectrophotometry has excellent spectroscopic performance suitable even for research applications as well as educational. The advanced optical design features a wide wavelength range of 190–1100 nm, stray light less than 0.02% and a spectral bandwidth of 1.0 nm, enough to satisfy any pharmacopeia requirement, and can performs spectral measurements at scanning speed up to 8000 nm/min [15]. The instrument developed for the electrical acquisition of the voltage and current of the solar panel has been subjected to a calibration process based on a digital multimeter tester DGM-360 [16]. This instrument has a resolution on the voltage of 0.01 V in the range of 60 V–4 V, and on the current intensity a resolution of 0.001 A in the measurement range 6 A. The uncertainty given by the manufacturer is 0.02% [12].

4. Electrical modeling of PV panel under dust

In the literature, there are several models describing the electrical behavior of a PV cell. The one-diode and two-diode models are widely used to obtain the I–V characteristic of the PV cell or panel output [17]. However, as shown in **Figure 4**, the one-diode model is the simplest one, moreover it is improved by incorporating a series resistor Rs [18, 19] and an additional shunt resistor Rsh [20].

The electrical current of the PV panel for the one diode model is given by:

$$I_{pv} = I_{ph} - I_0\left[exp\left(\frac{V_{pv} + R_s I_{pv}}{n_1 V_{th}}\right) - 1\right] - \frac{V_{pv} + R_s I_{pv}}{R_{sh}} \tag{1}$$

Where *Iph* is the photo-generated current dependent on *G* solar radiation, and temperature according to the relationship:

$$I_{ph} = \left(I_{ph,ref} + \alpha_0 .\Delta T\right)\frac{G_t}{G_{ref}} \tag{2}$$

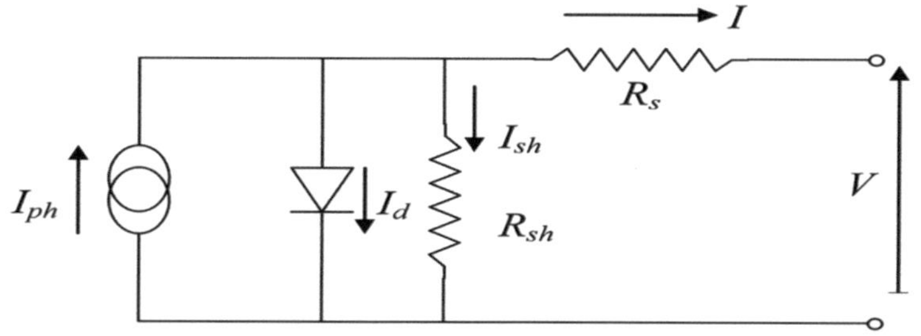

Figure 4.
One diode equivalent circuit of a PV cell.

With:
a_0 temperature correction coefficient for current (°C^{-1}).
$\Delta T = (T - Ta)$.
G solar irradiance on module plane (W/m^2).
G_{ref} = 1000 W/m^2.
$I_{ph,ref}$ photo-generated current (A).
Ipv solar cell terminal current (A).
I_0 reverse saturation current (A).
R_S series resistance (W).
R_{sh} shunt resistance (W).
STC standard test conditions (G_{ref} = 1000 W/m^2, T = 25°C and AM = 1.5).
T cell or module operating temperature (°C).
T_a ambient temperature (°C).
V_{pv} solar cell output voltage (V).
$V_{th} = \frac{kT}{q}$ thermal voltage (V).

Figure 5 shows the block diagram of the experimental I–V curve plotter for PV modules using the IRF740 MOSFET transistor as electronic method. When the control voltage V_{GS} is applied to its grid, it generates an output current Ipv variable quickly from 0 to Isc as well as a variable output voltage Vpv from Voc to 0 [21].

The Arduino board generates a continuous signal by the pin Vcc = 5 V, then amplified with a DC/DC converter type Boost XL6009E1, then this signal is injected into an RC filter (R = 440 Ω, C = 4700 mF). This also requires the resistance R_{DS} to evolve gradually [14]. The three MOSFET operating regimes that describe the relationship between I_D as a function of V_{GS} and V_{DS} are [16]:

Blocking regime ID = 0 A, if:

$$V_{GS} < V_{th} \tag{3}$$

Ohmic regime, $I_D = K\left[2(V_{GS} - Vth)V_{DS} - V_{DS}^2\right]$, if

$$V_{GS}\text{–}Vth > 0 \& V_{GS}\text{–}Vth > V_{DS} \tag{4}$$

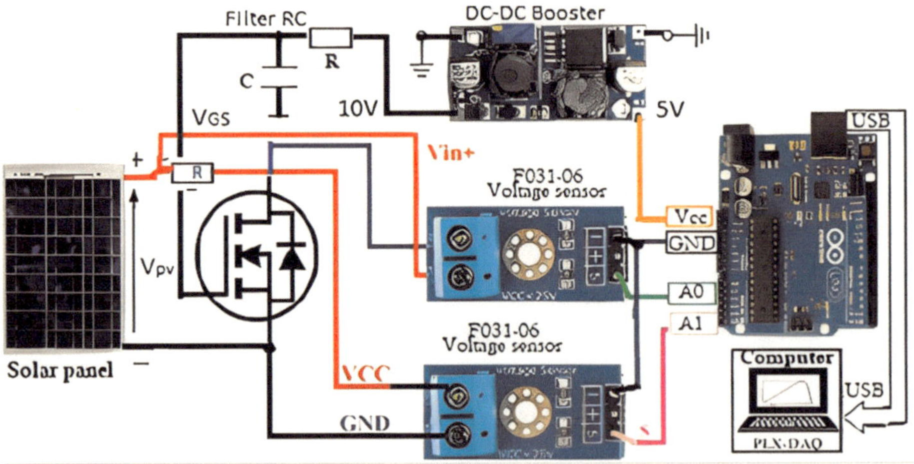

Figure 5.
Synoptic diagram of electronic charging technology.

Saturation regime, $I_D = K(V_{GS} - Vth)^2$, if

$$V_{GS}–Vth > 0\&V_{GS}–Vth < V_{DS} \quad (5)$$

Where K is the constant of the device and Vth the control threshold voltage of the transistor. By changing the value of V_{GS} within an appropriate range, the measurement points can vary between 0 and V_{OC}.

with:

$$I_D = I_{pv} \quad (6)$$

Arduino UNO is an open-source module based on a microcontroller, used to generate the PWM (Pulse Width Modulation) control signal. (PWM) is a technique used to control an analog circuit via a digital output, using the AnalogWrite function. This PWM signal is an electrical signal of maximum amplitude 5 V and constant frequency but with a variable duty cycle (**Figure 6**) [23].

The signal at the output of the DC/DC converter type Boost is injected into an RC filter, resistance R ≈ 110 W [24] and capacity of C ≈ 4700 mF [24], installed in series to vary the V_{GS} control voltage of the MOSFET, and finally trace the I-V & P-V characteristics of the solar PV panel.

According to **Figure 7**, the expression of the control voltage $V_{GS}(t)$ of the MOSFET is given by:

$$V_{GS}(t) = V_{PWM}.\left(1 - e^{-\frac{t}{RC}}\right) \quad (7)$$

Figure 6.
Boost DC/DC converter XL6009E1 [22].

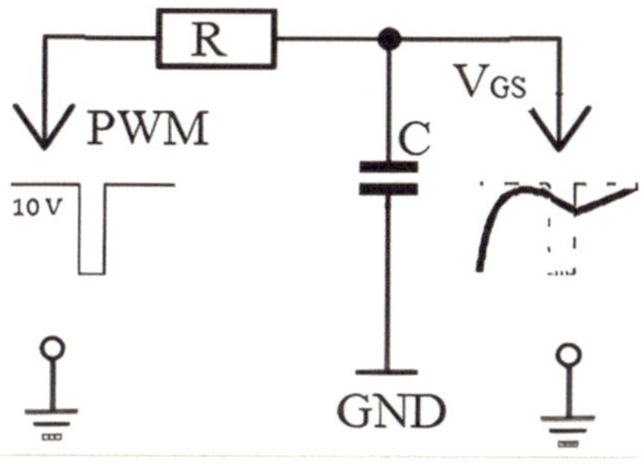

Figure 7.
RC filter circuit.

5. Simulation of the I-V characteristic with MATLAB/Simulink

To simulate the I-V characteristic of a PV panel, under MATLAB/Simulink using the MOSFET electronic load method shown in **Figure 8**.

block 1 in the previous **Figure 8** is to measure the current intensity in the circuit, block 2 a voltmeter, block 3 the PV panel (SX 330J), and block 4 represents the plotting of I-V and P-V curves, and finally block 5 represents the MOSFET used as a variable load controlled by block 6.

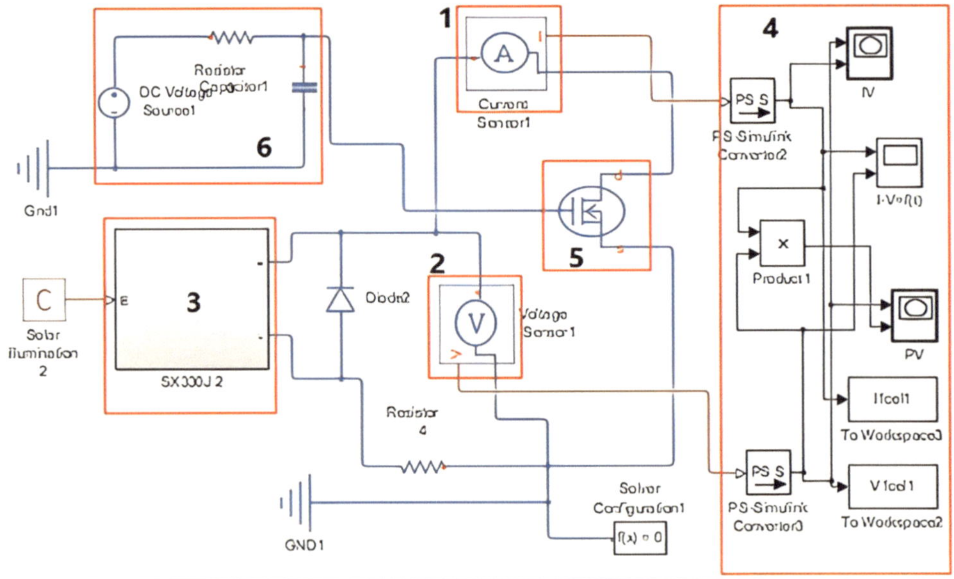

Figure 8.
Simulation circuit under Matlab/Simulink.

6. Results and discussion

6.1 Optical transmittance of (HP) and (AD) dust

Concerning the dust of the Agadir-Morocco site, the variations of the optical transmittance T(%) with the Wavelength, measured by the spectrophotometer JASCO V-730 for the samples dust collected from the two areas (HP and AD) are presented in **Figure 9**.

In this section, an amount of dust of 1.75 g/m^2, 3.45 g/m^2, 6.77 g/m^2, was collected from the two areas previously mentioned. The type of soil in the Adrar area (AD) is different from the type of soil in the Halieutic-Park area (HP). These results show the impact of soiling on the optical transmittance. A net drop off in the optical transmittance is observed depending on the origin of dust.

6.2 Electrical characterization and soiling effect

To understand the variation effect of the amount of the dust from (HP) to (AD) site, we considering the **Figure 10**, we found that for the same value of dust density,

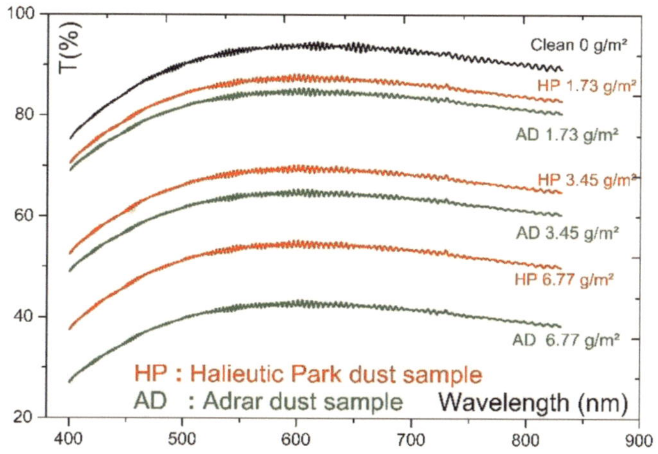

Figure 9.
Optical transmittance of (HP) and (AD) dust.

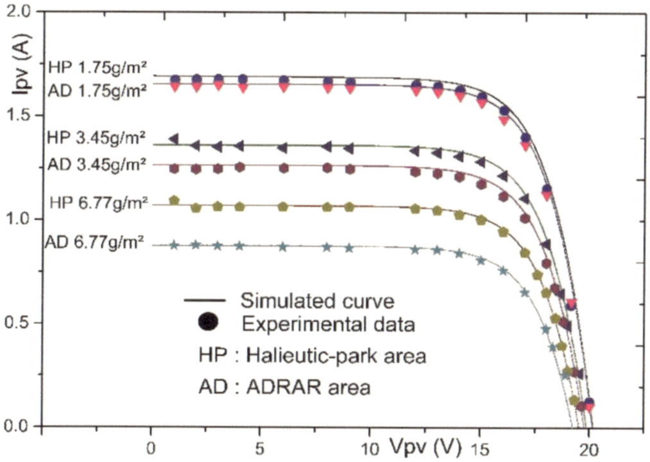

Figure 10.
I-V characteristics of SX-330 J panel.

the current intensity delivered by the PV panel, in the case of (HP) dust area, is higher than the current intensity delivered by the PV panel in the case of (AD) dust. In addition, the drop off in the density of the dust lead to a remarkable decrease in the current intensity. For both areas (HP and AD), in **Figure 11**, the calculated Er between experimental and simulated results is less than 15%.

The drop in power output caused by the accumulation of dust on the photovoltaic module surface is a big issue as reported by several authors [25–27]. **Figures 12–14** shows the evolution of the maximum power recorded on a PV module SX330J soiled at various dust densities: 1.75, 3.45, 6.77 g/m^2 (**Figure 15**).

The salient feature of our investigations is a clear decline of the maximum power output. Pmax decreases non-linearly with dust density. Our results are in good agreements with several works published by other authors [25, 26]. The maximum power Pmax of our solar panel drop off from 30 W to only 17 W when the dust density is

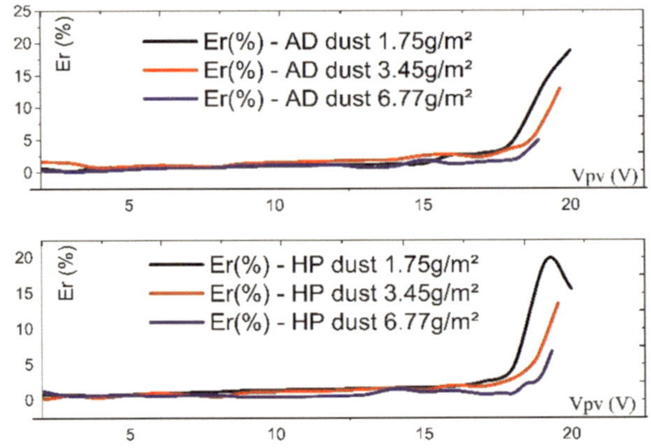

Figure 11.
Relatives error between experimental and simulated I-V curves.

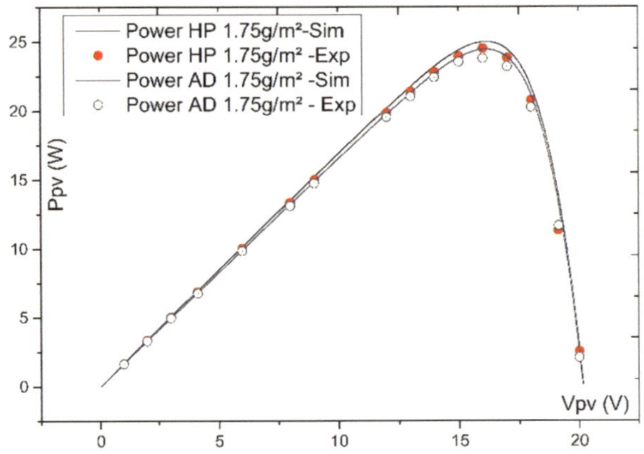

Figure 12.
P-V *curve, simulated and experimental under 1.75 g/m² of dust.*

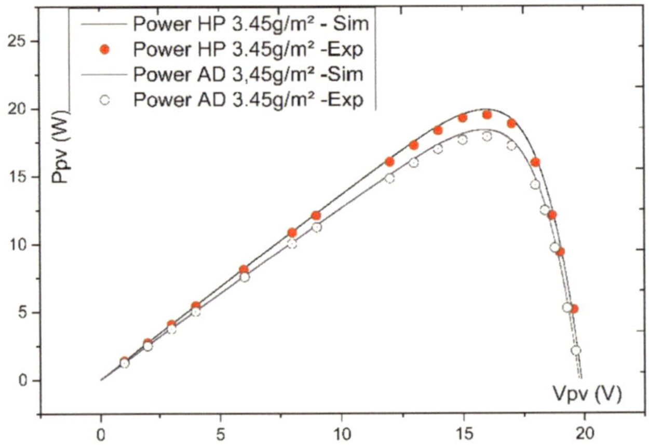

Figure 13.
P-V curve, simulated and experimental under 3.45 g/m² of dust.

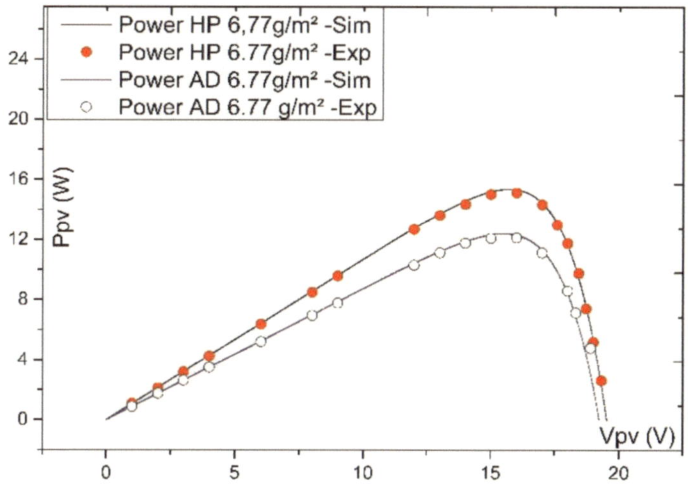

Figure 14.
P-V curve, simulated and experimental under 6.77 g/m² of dust.

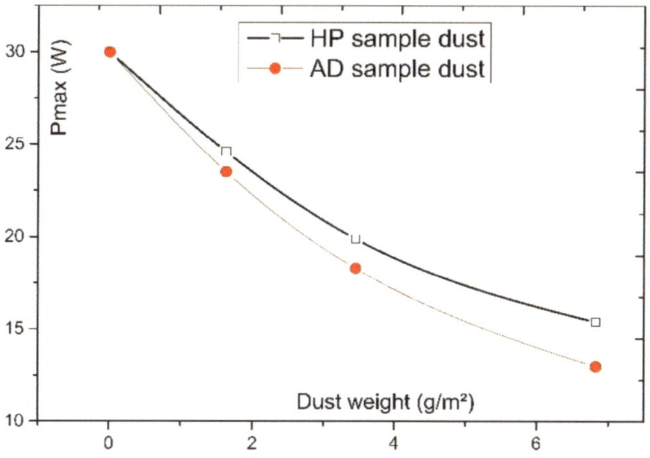

Figure 15.
Dust density effect on maximum power under dust type of AD and HP.

around 6.77 g/m^2 in the case of (HP) dust, and as weak as 14 W in the case of the (AD) dust. The observed behavior is correlated to the difference of dust distribution. Indeed, the size and distribution of the dust have a major effect on the degradation of PV performance, as small particles tend to block more sunlight radiation, with less space, and therefore contribute more to the deterioration of PV performance [28]. This can be explained by the fact that smaller particles are more uniformly distributed than larger particles [29], resulting in greater light scattering, especially at low intensities [30]. The huge loss (more than 50% in our case) of the generated PV power is a key factor of uncertainty and risk for solar production. As reported by several authors, less than 10% loss is the threshold for the need of cleaning. Such operation is inescapable for the recovery of the power [31, 32].

7. Conclusion

In the present chapter, for the first time, a comprehensive model was proposed for the PV module. This model has a very high strength in prediction of the electrical characteristics of the dusty PV module. The developed model was verified under different operating conditions of the PV module and then was used to predict the electrical performances of the dusty PV module in the region of Agadir, Morocco. The results show that this model can reasonably predict the module performances under different weather conditions. However, this study is still open as some other factors like humidity, wind speed, particle material, and PV module type need further investigation in the near future in order to approach a complete modeling. This study is highly useful in dusty regions and can help the PV designer to predict the power of the PV module with a reasonable accuracy.

Conflict of interest

The authors declare that there is no conflict of interest regarding the publication of this chapter.

Data availability

No data were used to support this study.

Author details

Abdellah Asbayou*, Lahoussine Bouhouch, Ismail Isknan and Ahmed Ihlal
Materials and Renewable Energies Laboratory, University Ibn Zohr, Agadir, Morocco

*Address all correspondence to: abdellah.asbayou@rdu.uiz.ac.ma

References

[1] Zhao Y. Fault Analysis in Solar Photovoltaic Arrays. Department of Electrical and Computer Engineering. Boston, Massachusetts: Northeastern University. Dec 2010. pp. 12-13

[2] Thirunavukkarasu GS, Seyedmahmoudian M, Chandran J, Stojcevski A, Subramanian M, Marnadu R, et al. Optimization of mono-crystalline silicon solar cell devices using PC1D simulation. Energies. 2021;**14**(16): 4986. DOI: 10.3390/en14164986

[3] Sarver T, Al-Qaraghuli A, Kazmerski LL. A comprehensive review of the impact of dust on the use of solar energy: History, investigations, results, literature, and mitigation approaches. Renewable and Sustainable Energy Reviews. 2013;**22**:698-733. DOI: 10.1016/j.rser.2012.12.065

[4] Costa S, Diniz AC, Kazmerski LL. Dust and soiling issues and impacts relating to solar energy systems: Literature review update for 2012-2015. Renewable and Sustainable Energy Reviews. 2016;**63**:33-61. DOI: 10.1016/j.rser.2016.04.059

[5] Javed W, Wubulikasimu Y, Figgis B, Guo B. Characterization of dust accumulated on photovoltaic panels in Doha, Qatar. Solar Energy. 2016; **142**(2017):123-135. DOI: 10.1016/J.SOLENER.2016.11.053

[6] Darwish ZA, Kazem HA, Sopian K, Al-Goul MA, Alawadhi H. Effect of dust pollutant type on photovoltaic performance. Renewable and Sustainable Energy Reviews. 2015;**41**: 735-744. DOI: 10.1016/j.rser.2014.08.068

[7] Adinoyi MJ, Said SAM. Effect of dust accumulation on the power outputs of solar photovoltaic modules. Renewable Energy. 2013;**60**:633–636. DOI: 10.1016/j.renene.2013.06.014

[8] Zorrilla-Casanova M, Piliougine J, Carretero M, Bernaola J, Carpena P, Mora-Lopez P, et al. Analysis of dust losses in photovoltaic modules. Photovoltaic Technology – Word Renewable Energy Congress. 2011;**2011**: 2985-2992

[9] Soteris PGK, Rafaela AA. On-site PV characterization and the effect of soiling on their performance. Energy. 2013;**51**: 439-446. DOI: 10.1016/j.energy.2012.12.018

[10] Said SAM, Hassan G, Walwil HM, Al-Aqeeli N. The effect of environmental factors and dust accumulation on photovoltaic modules and dust-accumulation mitigation strategies. Renewable and Sustainable Energy Reviews. 2017;**82**:743-760. DOI: 10.1016/j.rser.2017.09.042

[11] Al-Hasan AY. A new correlation for direct beam solar radiation received by photovoltaic panel with sand dust accumulated on its surface. Solar Energy. 1998;**63**(5):323-333. DOI: 10.1016/S0038-092X(98)00060-7

[12] Asbayou A, Aamoume A, Elyaqouti M, Ihlal A, Bouhouch L. Benchmarking study between capacitive and electronic load technic to track I-V and P-V of a solar panel. 2022;**12**(1): 102-113. DOI: 10.11591/ijece

[13] Nichols D. Arduino-based data acquisition into excel, LabVIEW, and MATLAB. Physics Teacher. 2017;**55**(4):226-227. DOI: 10.1119/1.4978720

[14] Asbayou A, Agdam M, Aamoume A, Soussi A, Ihlal A, Bouhouch L. Utilization of MOSFET transistor as an electronic load to trace I-V and P-V curve of a solar panel. Web of Science. 2021;**01021**:1-8

[15] JASCO V-730. Available from: https://www.jascofrance.fr/spectroscopie/uv-visible-nir/v-730/. 2022

[16] Handheld Digital Multimeter GDM-360. 2022. Available from: https://www.gwinstek.com/englobal/products/downloadSeriesDownNew/9914/718

[17] Maruska TD, Moustakas HP. Influence of the Wavelength of incident light on shunt conductance and fill factor in amorphous silicon solar cells. IEEE Trans. Electron Devices. 1984;**31**(5): 551-558

[18] Soon JJ, Goh ST, Maneuver N. Multi-Dimension Diode Photovoltaic Model for Different PV Cell Technologies2014. DOI: 10.1109/ISIE.2014.6865012

[19] Azzouzi M, Popescu D, Bouchahdane M. Modeling of electrical characteristics of photovoltaic cell considering single-diode model. Journal of Clean Energy Technologies. 2016; **4**(6):414-420. DOI: 10.18178/jocet.2016.4.6.323

[20] Rashel MR, Albino A, Veiga A, Ahmed MT, Tlemcani M, Goncalves TCF. Comparison of photovoltaic panel's standard and simplified models. In: 2016 International Conference for Students on Applied Engineering. ICSAE 2016; 2017. pp. 133-136. DOI: 10.1109/ICSAE.2016.7810175

[21] Zhu Y, Xiao W. A comprehensive review of topologies for photovoltaic I–V curve tracer. Solar Energy. 2020;**196**: 346-357. DOI: 10.1016/j.solener.2019.12.020

[22] XL6009E1. "XL6009E1 Adjustable DC-DC Boost Converter Module Datasheet". 2022. Available from: https://www.addicore.com/XL6009E1-Boost-Converter-p/ad456.htm, 2022

[23] Jo S, Lee C, Kim M, Ferracane J, Lee I. Effect of pulse-width-modulated LED light on the temperature change of composite in tooth cavities. Dental Materials. 2019;**35**(4):554-563. DOI: 10.1016/j.dental.2019.01.009

[24] Datasheet, "Datasheet. résistance et capacité". 2022. Available from: http://www.e44.com/composants/composants-passifs/resistances/

[25] Aïssa B, Isaifan RJ, Madhavan VE, Abdallah AA. Structural and physical properties of the dust particles in Qatar and their influence on the PV panel performance. Scientific Reports. 2016; **6**(1):1-12. DOI: 10.1038/srep31467

[26] Lopez-Garcia J, Pozza A, Sample T. Long-term soiling of silicon PV modules in a moderate subtropical climate. Solar Energy. 2016;**130**:174-183. DOI: 10.1016/j.solener.2016.02.025

[27] Seyedmahmoudian M, Mekhilef S, Rahmani R, Yusof R, Yusof ERR, Renani ET. Analytical modeling of partially shaded photovoltaic systems. Energies. 2013;**6**(1):128-144. DOI: 10.3390/en6010128

[28] Tanesab J, Parlevliet D, Whale J, Urmee T, Pryor T. The contribution of dust to performance degradation of PV modules in a temperate climate zone. Solar Energy. 2015;**120**:147-157. DOI: 10.1016/J.SOLENER.2015.06.052

[29] Hassan Qasem RG, Betts TR, Müllejans H, AlBusairi H. Dust-induced

shading on photovoltaic modules. Progress in Photovoltaics Research and Applications. 2012;**22**(2):218-226

[30] Sayyah A, Horenstein MN, Mazumder MK. Energy yield loss caused by dust deposition on photovoltaic panels. Solar Energy. 2014;**107**:576-604

[31] Pouladian-Kari A, Eslami S, Tadjik A, Kirchner L, Pouladian-Kari R. A novel solution for addressing the problem of soiling and improving performance of PV solar systems. Solar Energy. 2022;**241**:315-326

[32] Micheli L, Fernández EF, Aguilera JT, Almonacid F. Economics of seasonal photovoltaic soiling and cleaning optimization scenarios. Energy. 2021;**215**(A):119018

Chapter 8

Influence of Wind Incidence Angle on the Cooling of Rooftop-Mounted Solar Panels

Hadi Ahmadi Moghaddam, Matthew Phillips, Svetlana Tkachenko and Victoria Timchenko

Abstract

The cooling of PV panels is crucial because their electrical output and lifespan are adversely affected as their operating temperature rises. Considering wind current cooling impacts on the rooftop-mounted solar panels, adopting the local climate conditions such as dominant wind patterns is recommended to the building sector so that new buildings are placed considering the local wind directions. A 3D CFD model employing the URANS approach is developed to show the impacts of wind direction on the cooling rate of a PV panel installed on the surface of a slanted roof. The radiation effect is considered using the surface-to-surface radiation model. Two free stream velocities of 2 and 5 m/s and seven wind angles between 0 and 180 degrees are modelled. The results showed an optimum incidence angle at which the panel experiences lower temperatures. At wind angles below 90 degrees where there is direct contact between the wind flow and PV surfaces, the convective cooling rate is higher which in turn decreases the PV temperature. However, at higher angles, due to the presence of walls and edges of the structure, the wind flow is redirected resulting in the formation of wind flow separation. Therefore, convective cooling degrades, and PV experiences higher temperatures.

Keywords: solar panel, wind, convective cooling, CFD, wind incidence angle

1. Introduction

Due to the higher clean energy demands, many buildings nowadays are equipped with photovoltaic (PV) panels to convert solar radiation into electricity. While solar panels are generally adapted to the structure of a building in the form of façade-mounted and rooftop-mounted designs, the latter design is widespread. In the case of inclined roofs, the PV panels are installed at a distance on the surface of the roof. The gap between the panels and inclined roof surface is crucial to allow airflow to pass from the gap and cool down the PV from the backside, whether by passive cooling due to the formation of buoyant forces in the gap or forced convective cooling due to wind effects. Cooling of PV panels is crucial as their electrical efficiency and lifetime are adversely affected as they experience higher temperatures [1]. Research conducted by

IntechOpen

Ritzen et al. [2] on the performance of rooftop-mounted PV modules showed that the efficiency of non-ventilated modules degraded by 86% after 3 years.

Several studies have been conducted to appraise the performance of PV modules under wind or passive cooling effects. Chowdhury et al. [3] performed wind tunnel experiments and CFD simulations under different wind velocities and air gaps. Experiments by Mirzaei et al. [4] showed that higher wind velocities result in higher heat exchanges between the PV panel and airflow and consequently lower PV temperatures. They also suggested installing PV modules in a stepped open arrangement instead of a flat arrangement for better cooling. Lai and Hokoi [5] performed experimental and numerical studies on double-skin façade PVs and reported that the electrical efficiency of ventilated modules was higher by about 16–44% than the non-ventilated ones. Within another numerical research, Gan [6] concluded that adequate spacing or gap between the modules and the surface of the wall is required to allow the air to pass and cool down the PVs and avoid the formation of hot spots in panels.

The current numerical study investigates the impacts of wind direction on the cooling rate of a PV panel installed on the surface of a slanted roof. The slanted structure has been studied in this research as it is a common configuration used in Australia. In the following section, the studied case and the numerical procedure have been explained. Then, the results have been presented and discussed.

2. Case study

As shown in **Figure 1**, a 2.4 m height structure with a PV panel installed on its slanted roof has been studied numerically in this research. To evaluate the effects of wind direction, seven angles between the wind direction and the structure (0, 30, 60, 90, 120, 150, and 180 degrees) and two free stream velocities of 2 and 5 m/s have been considered, the main directions are shown in **Figure 1**. The distance between the panel and the roof surface is 0.1 m. The inclination angle of the roof or module is 30 degrees. The dimensions of the module were 1 m by 1.3 m, oriented such that 10 cm of the

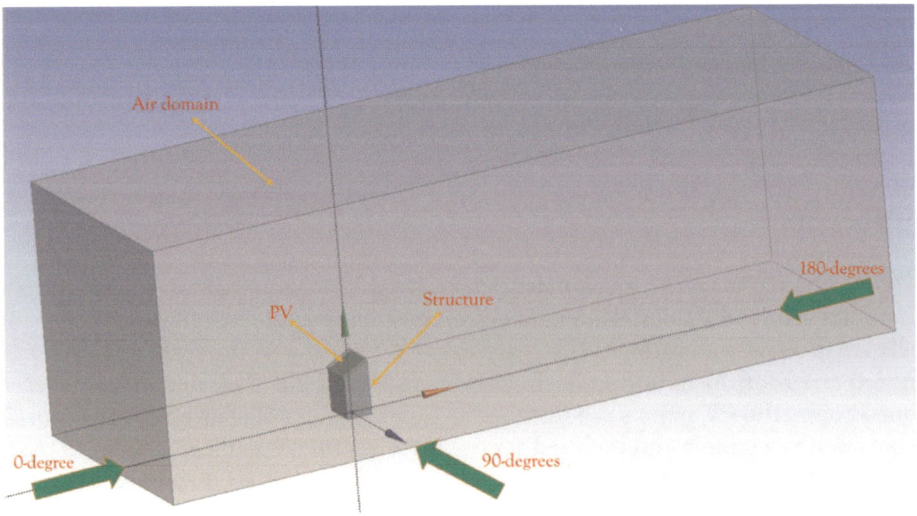

Figure 1.
Schematic view of the rooftop-mounted PV and air domain with wind directions.

Material	Thickness (m)	Thermal conductivity (W/m.K)	Heat capacity (J/kg.K)	Density (kg/m^3)
Glass	3000×10^{-6}	1.8	500	3000
EVA	500×10^{-6}	0.35	2090	960
Silicon	175×10^{-6}	130	700	2329
Tedlar	100×10^{-6}	0.2	1250	1200

Table 1.
Characteristics of the PV panel.

roof-top on all sides remained unobscured by the module when viewed from above. Adequate distance between the structure and the domain boundaries has been considered to correctly model the wind effects.

The following equation has been used for the wind velocity profile [7]:

$$U(h) = \left(\frac{h}{h_o}\right)^{\beta} U_{ref} \tag{1}$$

where β is a constant equal to 0.17, h_o is the total inlet height of 10 m, and U_{ref} is the average freestream velocity (2 or 5 m/s).

Further, the applied solar heat flux on the PV cells is 600 W/m^2. It is worth mentioning that of the total solar radiation incident upon a PV panel, a portion is reflected or converted into electricity (about 20%). The rest is converted into heat which in turn increases the PV temperature. Here, the applied heat flux of 600 W/m^2 is the amount of radiation that is converted into heat and modelled as heat generation in the simulation. The PV panel consists of five layers including glass, EVA, PV cells, EVA, and Tedlar, respectively, glass being the layer adjacent to the external domain and Tedlar adjacent to the air gap between the panel and roof. The characteristics of the PV panel are shown in **Table 1** [8].

3. Numerical setup

Unsteady Reynolds-Averaged Navier–Stokes (URANS) approach has been used to model both free and forced convective flows around the structure. For this purpose, Ansys Fluent 2021 code has been used. The working fluid (air) is considered to follow the ideal gas behaviour. The governing equations of the current problem, i.e., conservation of mass, momentum, and energy are:

$$\frac{\partial \rho}{\partial t} + \frac{\partial(\rho u_j)}{\partial x_j} = 0 \tag{2}$$

$$\frac{\partial(\rho u_i)}{\partial t} + \frac{\partial(\rho u_i u_j)}{\partial x_j} = -\frac{\partial p}{\partial x_i} + \frac{\partial \sigma_{ij}}{\partial x_j} + \left(\rho - \rho_{ref}\right) g_i \tag{3}$$

$$\frac{C_p \partial(\rho T)}{\partial t} + \frac{C_p \partial(\rho u_i T)}{\partial x_j} = k \frac{\partial}{\partial x_j}\left(\frac{\partial T}{\partial x_j}\right) + \frac{\partial q_j}{\partial x_j} \tag{4}$$

In Eqs. (2)–(4), g, ρ_{ref}, u, T, P, C_p, σ_{ij}, k, and q_j show the gravity vector, reference density, velocity vector, temperature, dynamic pressure, the specific heat capacity of constant pressure, stress tensor, thermal conductivity, and turbulent thermal flux vector, respectively.

To model the turbulent flow, the $k - \omega SST$ model has been used. SST is a hybrid model that utilises $k - \omega$ formulation within the boundary layer where viscous forces are dominant and then switches to $k - \varepsilon$ formulation outside the boundary layer making it suitable for a wide range of engineering applications [9]. A pressure-based solver along with a second-order discretization scheme has been used to solve the flow variables, and the coupled algorithm has been used for velocity–pressure coupling. The solution residual target is set to 10^{-5}.

An unstructured mesh with 2.06 million cells has been used to model the flow. It was found that a further increase in the number of cells did not affect the results. To correctly capture the flow behaviour within the boundary layer, inflation has been used on the solid boundaries. Generally, the mesh has been refined around the structure, within the gap between the panel and roof, and close to walls where a sharp gradient of flow variables such as velocity or temperature is expected.

The surface-to-surface radiation model has been used to simulate the radiation effects. This model uses the following formulation to compute the radiative energy transfer from surfaces:

$$q_{out,m} = \epsilon_m \alpha T^4 + \Omega_m \cdot \sum_{j=1}^{N} F_{jm} q_{out,j} \tag{5}$$

where $q_{out,m}$ is the energy flux leaving the surface m, α is the Stefan–Boltzmann constant, $\Omega_m = 1 - \epsilon_m$ is the reflectivity of surface m, ϵ_m is the emissivity of the surface m, and F_{jm} is the view factor between surface m and surface j.

4. Results and discussion

Prior to presenting the results, it should be noted that all presented results are averaged over time. The change of the mean temperature of PV with wind direction is shown in **Figure 2**. It is seen that for both wind velocities of 2 and 5 m/s, the mean temperature of the PV is lower at angles below 90 degrees which is associated with higher surface heat transfer coefficients, as shown in **Figure 3**. Under these conditions, there is direct contact between the wind and module surfaces which contributes to forced convective cooling of the PV, thereby increasing the surface heat transfer coefficient, and decreasing the temperature of PV. At lower angles, the wind penetrates the gap between the PV and roof surface resulting in an increased heat transfer. In other words, the forced convective cooling effects on both sides (front and back) of the module are prominent at angles below 90 degrees. It is noticeable that by increasing the angle from 0 to 60 degrees, there is a slight decrease in the mean temperature of PV. This is mainly because, at angles higher than 0 degrees (and below 90 degrees), more airflow penetrates the gap between the PV and roof which in turn intensifies the convective cooling effects from the backside of the PV. **Figure 4** shows the schematic view of the airflow penetrating the gap between the PV module and roof for the wind angles of 0 and 60 degrees. As seen, when the wind angle is 0 degrees, the airflow mainly enters the gap from the bottom side. However, for the 60 degrees case, the

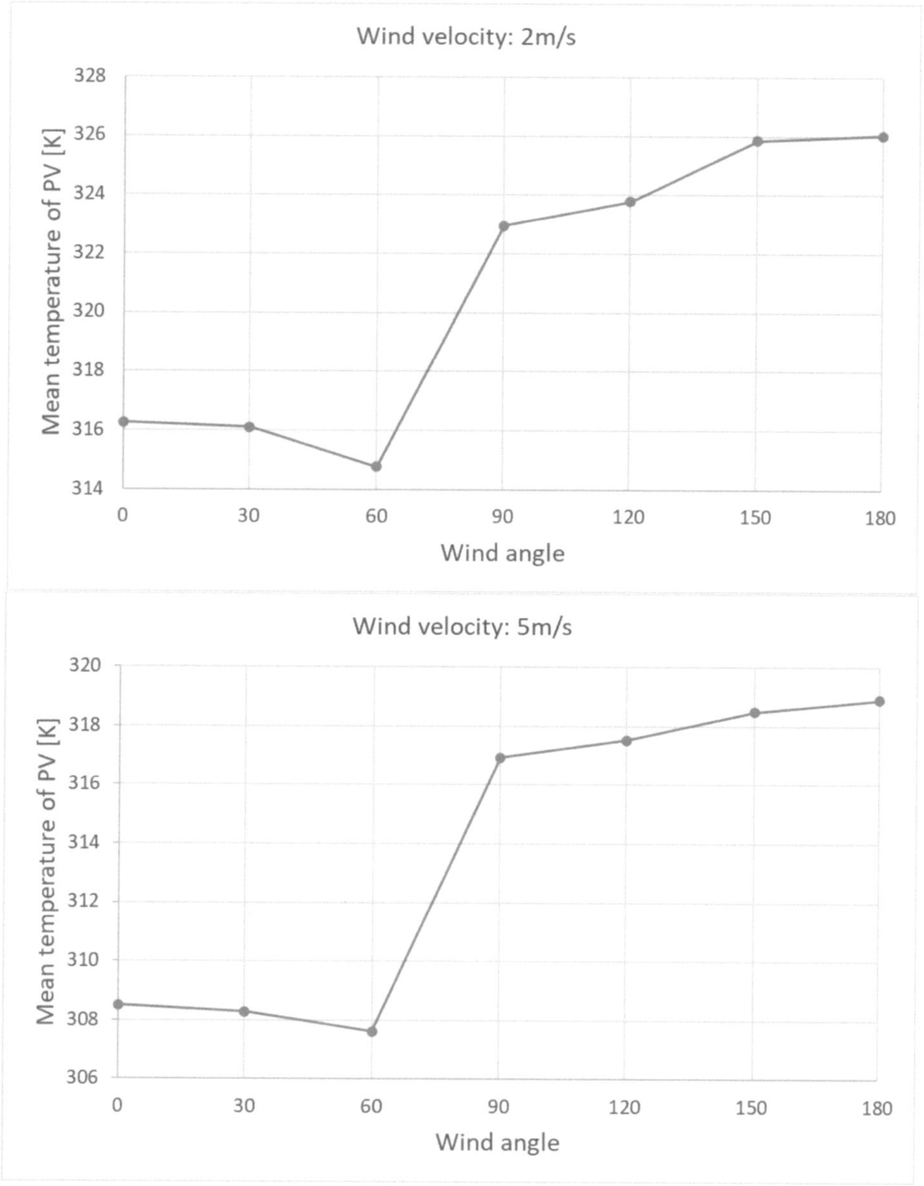

Figure 2.
Effects of wind angle on the mean temperature of PV.

airflow enters the gaps mainly from both bottom and left side of the gap, thereby intensifying convective cooling effects.

It is worth noting that similar observations were reported by the experiments of Wen [10] who studied the cooling effects of wind direction on a PV panel whose surface was parallel to the wind and observed that the convective heat transfer coefficient reached a maximum value for the angles in the range of 60–120 degrees depending on the wind velocity. However, as will be discussed in the following, the HTC drops drastically in the current research for angles beyond 60 degrees due to the edges of the structure that act as a barrier.

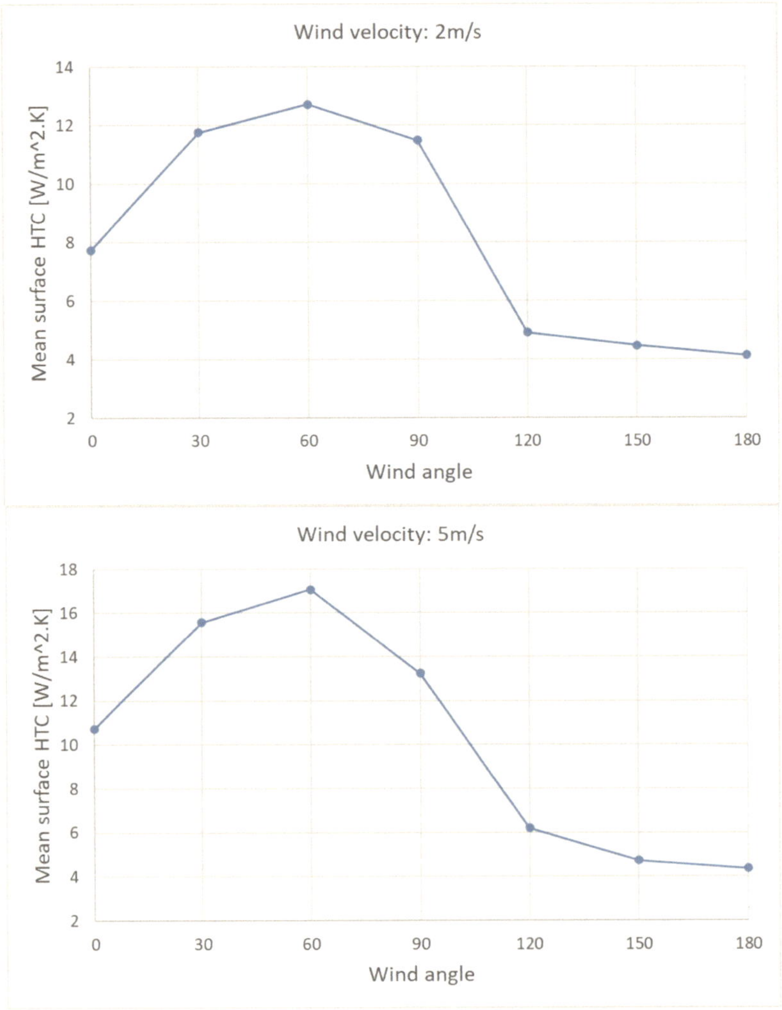

Figure 3.
Effects of wind angle on the mean surface heat transfer coefficient (HTC) of PV.

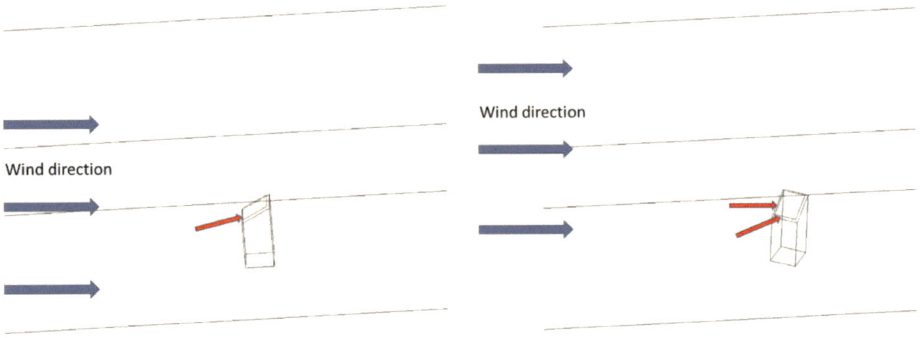

Figure 4.
Schematic view of the airflow penetrating the gap between the PV and roof, left) 0 degrees and right) 60 degrees wind angle.

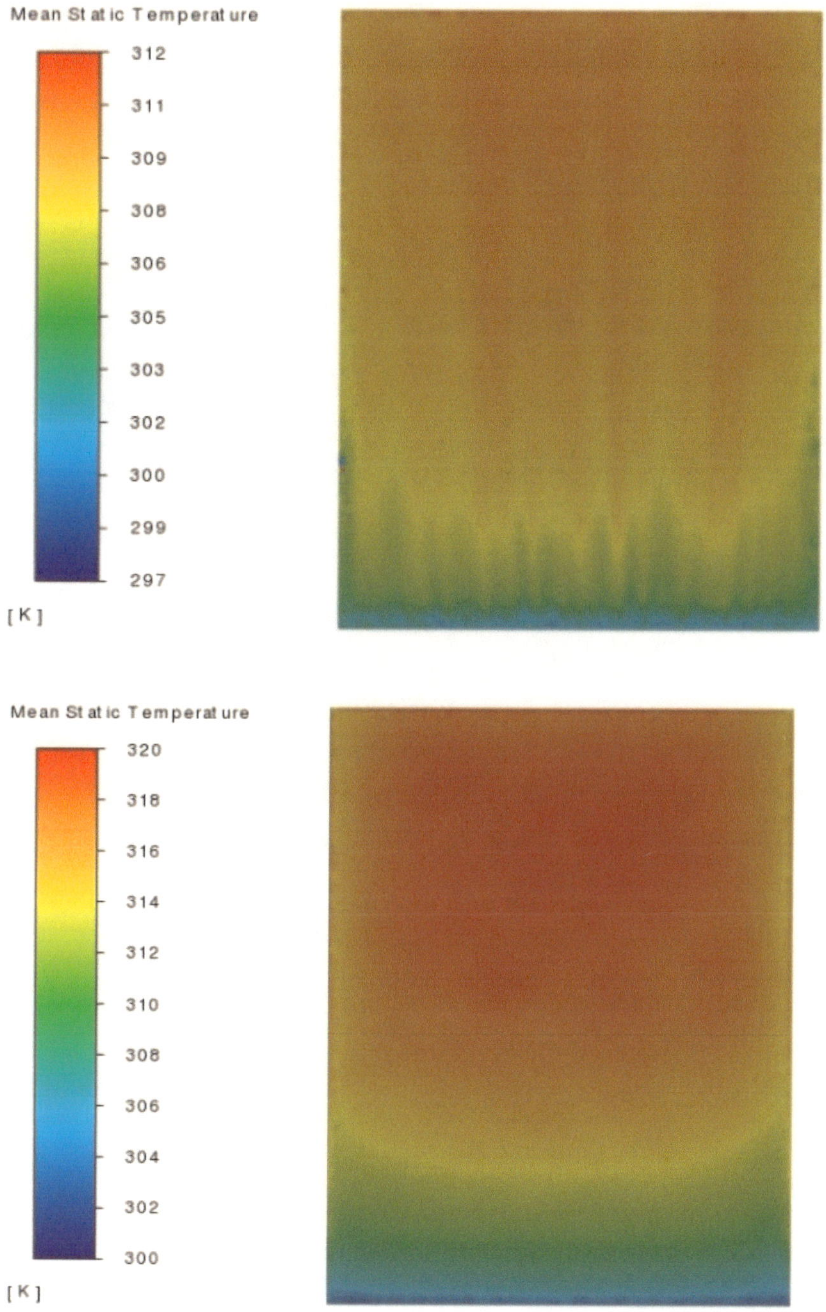

Figure 5.
Contours of mean static temperature of PV panel for free stream velocity of 5 m/s, top) 0 degrees, bottom) 180 degrees.

By increasing the wind angle beyond 60 degrees, there is a sharp increase in the mean temperature and a decrease in the mean surface HTC of the PV, as shown in **Figures 2** and **3**, respectively. The reason behind this is that at higher angles, the

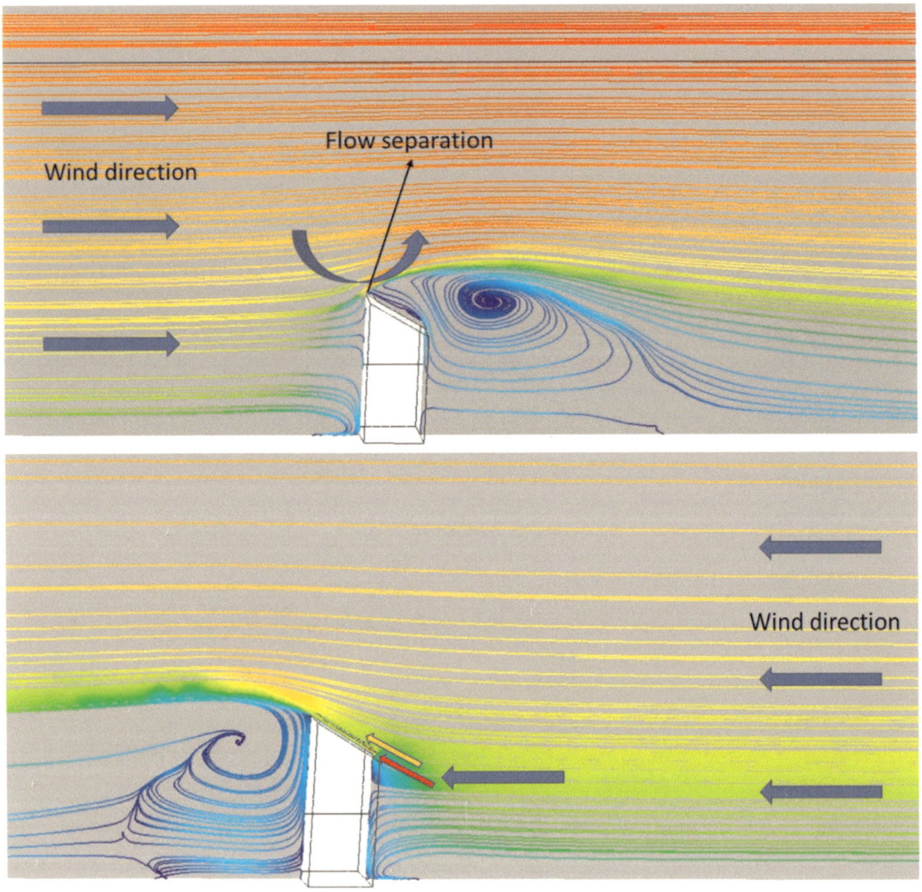

Figure 6.
Flow streamlines for wind velocity of 5 m/s, top) 180 degrees, bottom) 0 degree.

inclined roof and sidewalls of the structure act as a barrier between the module and wind, thereby decreasing the direct contact between the wind and module and consequently degrading the forced convective cooling effects.

Figure 5 illustrates the temperature contours of the PV panel for wind angles of 0 and 180 degrees. As observed, PV experiences lower temperature due to higher cooling effects at zero wind angle. It is also observed in **Figures 2** and **3** that, as expected, mean HTC augments and mean temperature of PV decreases as the wind velocity increases due to the enhanced forced convective cooling effects.

To get a better understanding of the wind and structure interactions, **Figure 6** is presented showing the streamlines for the wind velocity of 5 m/s and angles of 0 and 180 degrees. As seen, for the wind angle of zero degrees, the wind flow approaching the PV is divided into two streams, one passing beneath the PV and the second passing over the surface of the PV resulting in convective cooling effects from both sides. A large vortex is being also generated behind the structure. However, at the angle of 180 degrees, the walls and edges of the structure result in the separation of flow and formation of a large vortex in the vicinity of PV which weakens the convective cooling rates.

5. Conclusion

A 3D transient CFD model was developed to illustrate the impacts of wind direction on the cooling of a PV panel installed on the surface of a slanted roof. The simulation includes radiative effects by surface-to-surface radiation model, natural convection generated by the heated PV due to the buoyancy effects and forced convection due to the wind effects. The numerical results showed that the wind direction has a significant impact on the cooling rates of the module. At angles below 90 degrees, due to direct contact between the wind and sides of the PV panel, the cooling rate of the PV is higher resulting in lower PV temperatures. However, at higher angles, the walls and edges of the structure act as a barrier between the wind flow and PV and prevent direct contact between them. Therefore, the convective cooling rates degrade, and PV temperature rises.

It should be noted that the results may depend on other geometrical parameters such as distances from the edges of the PV module to the edges of the roof and building height, which could be investigated in future studies.

Acknowledgements

The funding provided for this research through the Australian Government Research Training Program (RTP) Scholarship is appreciated. This research includes computations using the computational cluster Katana supported by Research Technology Services at UNSW Sydney.

Author details

Hadi Ahmadi Moghaddam*, Matthew Phillips, Svetlana Tkachenko and Victoria Timchenko
School of Mechanical and Manufacturing Engineering, UNSW, Sydney, Australia

*Address all correspondence to: h.ahmadi_moghaddam_dastjerdi@unsw.edu.au

References

[1] Tkachenko SA et al. Computational study of natural convection flow in an open-ended channel coupled with a room: Application to building-integrated photovoltaic (BIPV) systems. In: Proceedings of CHT-21 ICHMT International Symposium on Advances in Computational Heat Transfer. Begel House Inc.; 2021

[2] Ritzen M, Vroon Z, Rovers R, Geurts C. Comparative performance assessment of a non-ventilated and ventilated BIPV rooftop configurations in the Netherlands. Solar Energy. 2017; **146**:389-400

[3] Chowdhury MG, Goossens D, Goverde H, Catthoor F. Experimentally validated CFD simulations predicting wind effects on photovoltaic modules mounted on inclined surfaces. Sustainable Energy Technologies and Assessments. 2018;**30**:201-208

[4] Mirzaei PA, Paterna E, Carmeliet J. Investigation of the role of cavity airflow on the performance of building-integrated photovoltaic panels. Solar Energy. 2014;**107**:510-522

[5] Lai C-M, Hokoi S. Experimental and numerical studies on the thermal performance of ventilated BIPV curtain walls. Indoor and Built Environment. 2017;**26**(9):1243-1256

[6] Gan G. Numerical determination of adequate air gaps for building-integrated photovoltaics. Solar Energy. 2009;**83**(8): 1253-1273

[7] Elshaer A, Aboshosha H, Bitsuamlak G, El Damatty A, Dagnew A. LES evaluation of wind-induced responses for an isolated and a surrounded tall building. Engineering Structures. 2016;**115**:179-195

[8] Zhou Z et al. Passive PV module cooling under free convection through vortex generators. Renewable Energy. 2022;**190**:319-329

[9] Moghaddam HA, Shafaee M, Riazi R. Numerical investigation of a refrigeration ejector: Effects of environment-friendly refrigerants and geometry of the ejector mixing chamber. European Journal of Sustainable Development Research. 2019;**3**(3): em0090

[10] Wen L. An investigation of the effect of wind cooling on photovoltaic arrays, JPL Publication 82-28, JPL Document No. 5101-201. DOE JPL-1012-69, Jet Propulsion Laboratory, Pasadena, California. 1982

Chapter 9

Performance Evaluation of Solar Cells by Different Simulating Softwares

Abdul Shakoor, Ghazi Aman Nowsherwan, Muhammad Fasih Aamir, Ahmar Ali, Sami Ur Rehman, Waheed Alam, Muhammad Yasir, Khizra Arif, Muhammad Ahmad and Jamal Yousaf

Abstract

In the contemporary era of technological advancements, solar energy emerges as a promising and easily implementable solution to meet future energy demands sustainably. This chapter delves into recent innovative techniques and simulation software pertaining to this environmentally friendly technology, focusing on device simulation, novel structures, and cutting-edge methods. A comparative analysis among major solar cell modeling simulators, such as PC1D, SCAPS-1D, wxAMPS-1D, AMPS-1D, ASA, Gpvdm, SETFOS, PECSIM, ASPIN, ADEPT, AFORS-HET, TCAD, and SILVACO ALTAS, is presented. These simulators not only aid in analyzing fabricated cells but also predict the impact of device modifications. The current year has witnessed significant efforts in developing sustainable energy systems through innovative solar cell simulators and semiconductor models. A concise evaluation of well-established solar cell simulators is provided to identify the most reliable tool for assessing photovoltaic technology performance. The chapter offers a user-friendly linear operating procedure and a congenial dialog box for multi-junction solar cells, providing valuable benefits for scientists, researchers, and skilled programmers in the photovoltaic community. This solar simulation software plays a crucial role in designing environment-friendly solar energy systems and calculating potential solar PV system outcomes for various projects, both grid-tied and off-grid, continually improving the solar energy technology landscape.

Keywords: solar cells, simulation software, performance evaluation, photovoltaic system, simulators

1. Introduction

Among different energy resources, sun's energy is introduced as a versatile source of energy which may be used for number of purposes like heating, cooling and brightening of houses and businesses. Sunlight is environment friendly source of energy that is converted into other useful form of energies for multiple purposes from different ways. The primitive solar designs for space heating, solar water heating and cooling are the most common solar applications for domestic and commercial use in present time. For solar designers and installers, it is important and critical to generate energy-efficient, environment friendly and cost-effective systems for both domestic and industry sectors. Due to these reasons, solar simulation software is important for researchers, engineers, solar cell designers, planners and dealers in market to create realistic offers for reliable attractive potential solar customers for sustainable future. Hence, let us get insight for some vital and novel information with respect to solar simulation software and informative knowledge about what are their use, how to choose one along with main phases in PV system design and its application in numerous fields.

Now days, for theoretical evaluation and investigations of solar cell working before final solar cell fabrication through different fabrication and deposition techniques, it is necessary for researchers to work on computer-based simulation programs or simulators that are becoming increasingly famous for their reliable theoretical assessment. The development of different solar simulators for photovoltaic applications was introduced in the mid-1980s when the technology was at its peak [1, 2]. With the passage of time, the increase in speed, performance and capacity of computers also enhanced the performance, speed and other features of the solar simulators. Despite the fact that fundamental objective for the invention of these simulating programs was to aid the researchers to reveal the well device performance to the academians but after some time it has grown into an essential tool in lab as well as for domestic and industrial applications. Up to now, several solver packages designed for simulating and modeling of solar cells are commercially available [3].

In the family of modern solar technology, photovoltaic community is familiar with various one-dimensional modeling tools that are mainly used like PC-1D, SCAPS-1D, AMPS-1D along with multi-dimensional modeling tools that are mainly used like Silvaco ATLAS, COMSOL multiphysics, Sentaurus ASPIN3 [4]. Generally, the basic modules of many simulators have similarities but they differ by features like speed, effectiveness, graphical interface and user access. PV simulators are usually designed and generated for various areas such as optics, semiconductor physics and electrical circuits along with characterization, production, costs analysis and photovoltaic operations. To produce ever-better materials and techniques for solar PV cells, there are numbers of parameters that may be accommodated for the improvement and enhancement of device performance along with multiple factors such as material type, geometric arrangement and thickness. The development of new solar technology is a challenging process that involves making small changes to various parameters. However, computational modeling simulators have made it easier and more accessible to evaluate and examine these advances, thereby eliminating the need to physically build every new change for evaluation. In this chapter, there are number of scientific publications on solar simulators [5, 6] and researchers mainly focused on modeling softwares of PV system and along with analysis and vice versa. In our work we have studied frequently used well-established solar simulators in details. We also reviewed the benefits and drawbacks of these tools, as well as comparisons with our prepared list of free simulation programmes.

2. Uses of solar simulation Softwares

According to the need and demand of solar energy, solar simulation software is used to build and model PV solar applications. They are used to assess the performance of PV systems. It aids in system design and fabrication process by evaluating the size specifications and choices of different solar power system components like solar panel array, charge controller, PV inverter and battery bank in addition to identify any losses in system. Solar simulator also calculates the impact of varying orientations that tilt angles on system working. The capacity to determine the cost as well as payback time period of domestic and industrial projects is a critical aspect of solar simulation software. Essentially, solar simulation software assists engineers in creating efficient but cost-effective PV systems.

3. Benefits of using solar simulation Softwares

Solar simulation software enables energy experts and designers to do a wide range of jobs and calculations with ease. Without them, these calculations would have taken a long time. It also provides easy automation, allowing solar community to provide consumer's rapid feedback. Solar simulation software simplifies and streamlines the engineering and sales processes, which is remarkably beneficial for large-scale solar production plants with different challenges.

4. Features of solar simulation Softwares

A simulator that possesses the following features should be considered a good simulator.

- Accuracy: The solar modeling programme must be precise in calculating many parameters like energy yield, inverter size, PV modules and vice versa.

- Ease of Use: The unique property of effective solar simulator is its easy use, which should be streamlined for both researchers and technicians to evaluate and operate.

- User-Friendly: The simulating software used to generate PV systems should be easy and simple for users so that they quickly obtain the required results after providing the input data.

- Flexibility: Another crucial and important feature that affects the use of a photovoltaic system designing tool. It helps in finding the size, thickness, quantity and also type of each necessary component used in the PV system, which results in better performance.

- Compatibility: Solar simulator should be compatible with vast operating systems like Windows, Linux, and Mac or any web browser.

- Report Generation: A good solar modeling software generates comprehensive results that help in evaluating the components required for the fabrication of a solar power system, also assist in determining and evaluating system energy losses, expenses and saving parameters. The generating report also expresses project details like project name, description, location and so on.

5. Key steps in designing a solar PV system

A photovoltaic system is designed in four major stages:

- System Design
- Size of component
- Performance Study
- Economic feasibility

- System Design: Gathering site location data is an important part of system design. It includes solar irradiation, utility connection and shading analysis needed in this data.
- Sizing: The secondary important phase is the sizing parameters that determine the size of various components of the PV system. Other features like battery storage capacity and inverter capacity are also mentioned. This ensures that required electricity is produced to fulfill energy demand.
- Performance Analysis: The solar system is simulated by using solar PV modeling software during the performance study.
- Economic Viability: The economic feasibility of the required project must be acknowledged. A brief economic and feasible project of the solar PV system should be performed at this fact.

6. Solar cell simulators

This chapter provides a detailed exploration of solar cell simulators, covering their development procedures, programming languages, semiconductor models and important features from the inception to the state-of-the-art versions. These modeling programs used commonly within the photovoltaic community in this modern era.

6.1 Personal computer simulation software (PC1D)

PC1D is the solar cell modeling application that is utilized the most frequently among those that are available for commercial purchase. The commercial success of this product is built upon its swiftness, user-friendliness and continuous updates to keep pace with the most recent cell models. It is used to model the performance of new devices and also helps new users build a grasp of the physics behind device operation. The University of New South Wales is now making PC1D freely accessible to the public. When the user has finished configuring the PC1D basic model, the system will create a number of nodes for them to solve. The number of nodes increases in regions of the cell when there is a change in the doping as well as in regions that are close to surfaces. It is also possible to halt the software during the simulation so that you may investigate the spatial distribution of carriers or the field throughout the device at a certain bias point. The schematic representation and doping density profile of Si-based solar is illustrated in **Figure 1**.

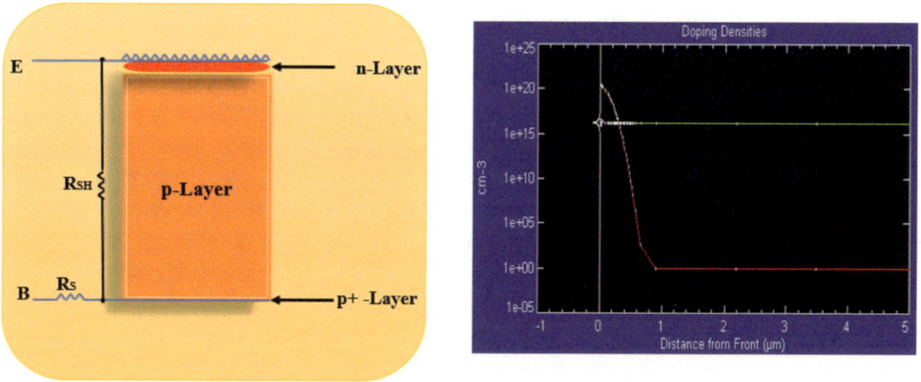

Figure 1.
Schematic representation of device architecture designed by PC1D and doping density profile generated by PC1D.

PC1D is known as one-dimensional solar modeling software generated by computer engineer Diane Rover et al. in 1985. It was installed at IBM-compatible personal computers and written in the Pascal language as well as designed for crystalline silicon solar cell [1, 2]. The exceptional parameters like speed, interface and continuous update to the up-to-date cell model use permit it for the best utilized package. For the improvement of the basic version of this simulator the mentioned models or methods were used such as the trap-assisted tunneling model, intra-band and Newton-Gummel method. On the other hand, the updated version, (PC1D mod 6.1), incorporates several new models like Fermi-Dirac statistics. In addition, crystalline silicon solar cell efficiency simulation is included for better executions [7]. In recent times, 20.35% efficiency has been achieved and simulated for single c-Si solar cell with the help of PC1D [8]. In addition with efficiency simulations, other factors like band gap's impact and electron affinity tuning of Zinc oxide layer on crystalline solar cell enactment as well as the factor like Anti-Reflecting Coating (ARC) layer's concerns on this solar cell have simulated by using this simulator for better performance [9].

6.2 Advanced semiconductor analysis (ASA)

Advanced Semiconductor Analysis (ASA) is a state-of-the-art software used for simulating thin-film hydrogenated solar cells. It is highly effective and widely regarded as the most advanced operational tool for this purpose. ASA solves one-dimensional semiconductor equations using free electron concentration, hole concentration and electrostatic potential (the Poisson equation and two continuity equations for electrons and holes). It also uses advanced physical models to characterize device and material optoelectronic characteristics. The updated version of this simulating package uses an integrated optoelectronic attitude for rapid simulations of JV curves, efficiencies and fill factors for microcrystalline and thin-film silicon solar cells [10, 11].

6.3 Amps-1D

AMPS-1D is another widespread simulation software for analyzing the characteristic performance of a-Si:H, polycrystalline, copper indium gallium selenide

[12–14] and Copper zinc tin sulfide solar cells. AMPS explains how material properties (bandgap, affinity, doping, mobilities, gap state defect distributions in the bulk and at interfaces) and device design/structure control device physics and response to light, impressed voltage and temperature. AMPS allows users to study and compare band diagrams, current components, recombination, generation and electric field plots as a function of light intensity, voltage, temperature and location to better understand device behavior to a given circumstance (i.e., light bias, voltage bias and temperature). AMPS-1D was supported by IBM and the Electric Power Research Institute [15]. It uses the FORTRAN programming language and is based on the Newton–Raphson method. Compared to earlier software, it has better simulation features. However, data entry in AMPS can be time-consuming due to the requirement of numerous parameters and layers. **Figure 2** shows parametric information that need to be added for single layer in AMPS-1D software.

6.4 WxAMPS

wxAMPS is a 1D solar cell modeling application developed in partnership with the University of Illinois at Urbana-Champaign and Nankai University in China. In 2012, Yiming Liu et al. developed well-run, up-to-date version of AMPS solar simulator [16]. This modified version takes the identical dataset as original AMPS-1D takes and confirms the similar type of physical values of defects and recombination. This tool is supplementary for tandem solar cells simulation created by alteration of a trap-assisted tunneling model [16]. Until now, this simulator is normally employed for simulation of CIGS solar cell, CZTS, dye sensitized solar cell (DSSC) [17–19] as well as tandem solar cell and amorphous silicon p-i-n tandem cell (**Figure 3**) [20].

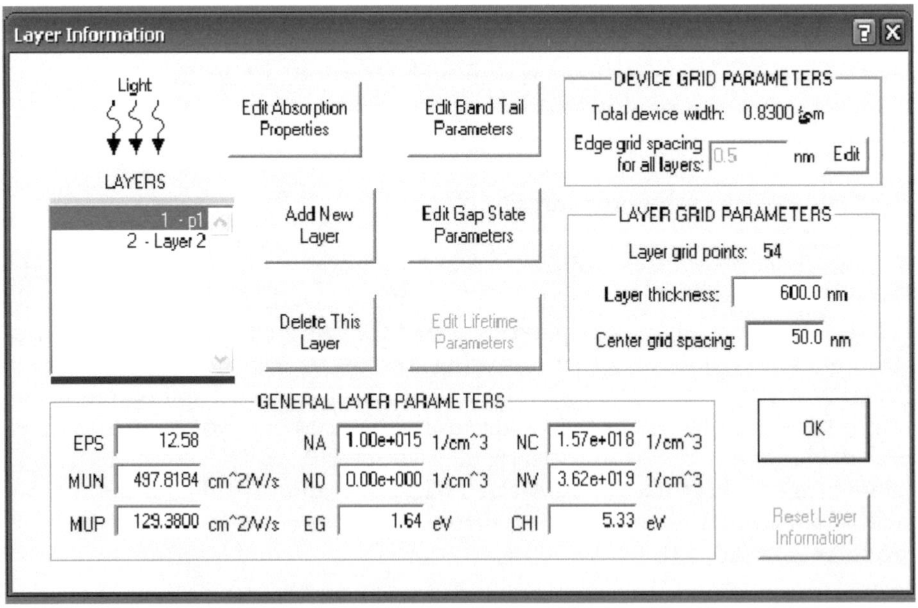

Figure 2.
Parametric configuration in AMPS simulation software.

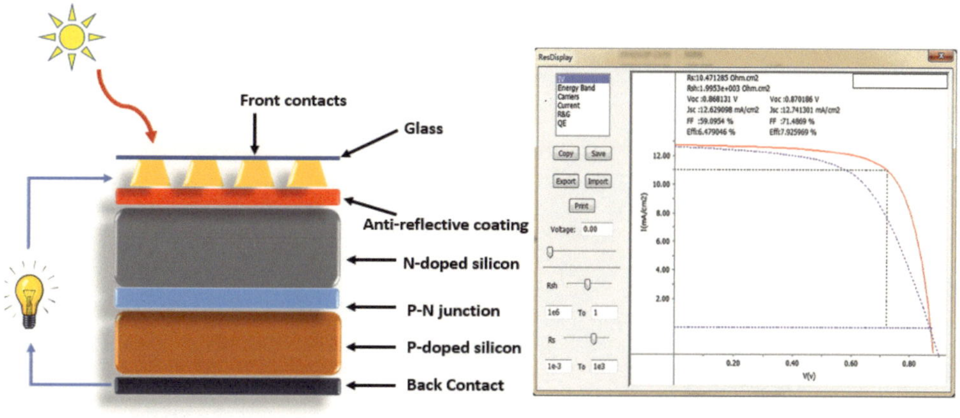

Figure 3.
Schematic representation of silicon-based solar cell designed by wxAMPS and input/output display panel wxAMPS [20].

6.5 SCAPS-1D

SCAPS-1D is a one-dimensional solar cell simulator commonly used for CIGS solar cell. It was developed by the Department of Electronics and Information Systems (ELIS) at the University of Gent, Belgium [21]. The simulation tool is freely accessible to the research community and is designed for polycrystalline cell structures of both CuInSe2 and CdTe communities. This simulating tool is developed to manage the following factors like thin-films, multiple interfaces and large band gaps etc. The software progressed over the years to comprise further mechanisms like Auger recombination, multiple enhancements to user interface, tunneling etc. This simulating tool was specially established with the help of Gummel scheme along with Newton Raphson method for CIGS as well as CdTe solar cell. Until now, this package has been operational for crystalline solar cells, GaAs and a-Si: H cell, as well as micro amorphous Si solar cells and their applications. The step-by-step procedure for numerical calculation and definition of panel for different layers employed in the SCAPS-1D device model is represented in **Figure 4**. Following features have given this tool a numerous identity [22].

- Data analysis and calculations for I-V, C-V and C-f.
- A number of standard models offered with the distribution packages.
- Well-established user interface, appropriate scripting facilities.

6.6 SETFOS

A CPU is an efficient and a powerful simulation software that is commercially available and designed for the development of unique optoelectronic technologies like organic thin-film-based PV solar cell, OLED and perovskite solar cells. This star software is generated by Professor Dr. Beat Ruhstaller [23]. There are four different modules in SETFOS for simulation given in following parameters [24].

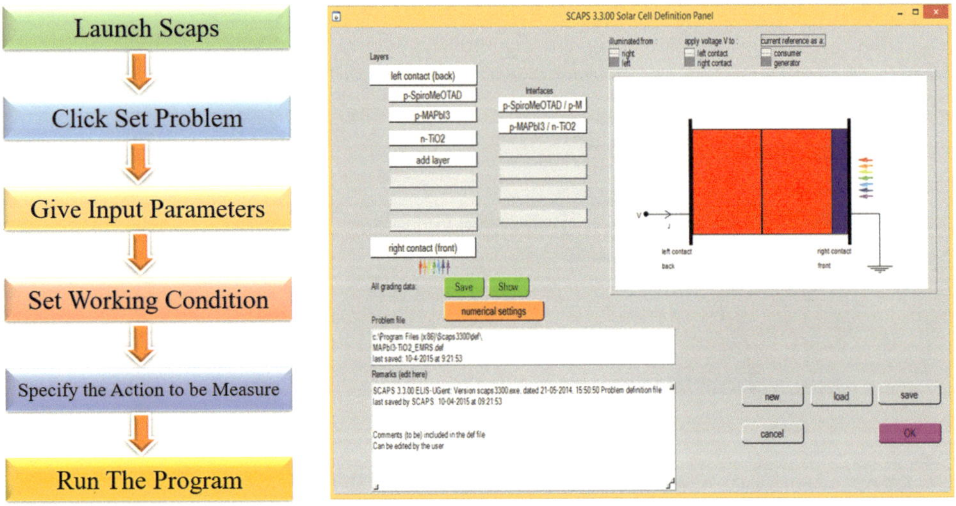

Figure 4.
Algorithm and definition of panel in SCAPS-1D.

- Light emission and absorption.
- Charge transport characteristics.
- Scattering.

This simulating tool has been declared as reliable and suitable tool for the development of organic solar cell device structure. The reliability, flexibility and speed of this software make it reliable and effective for perovskite solar cell, quantum dot and organic thin-film based solar cell (**Figure 5**).

6.7 General-purpose photovoltaic device model (Gpvdm)

GPVDM is a free and versatile simulating software for optoelectronic devices and was created by Roderick C. I. Mackenzie at Imperial College London. This tool was written for simulation of PV solar cells but later on it is proficient to simulate the working of multiple classes of devices such as OFETs, OLEDs, optical filters, first, second and third generation PV solar cells. Until now, this tool can simulate a lot of devices as mentioned below.

- Organic LEDs and Organic Solar Cells.
- Organic Field Effect Transistors (OFETs).
- Crystalline Silicon, CIGS solar cells and a-Si solar cells.

This simulating software has both electrical and optical modules for producing and analyzing the precise solar cell simulations for various applications with respect to requirements [25]. This platform assists an easy and operational tool to explore the effect of optoelectronic properties. **Figure 6** visualizes the graphical user interface in GPVDM.

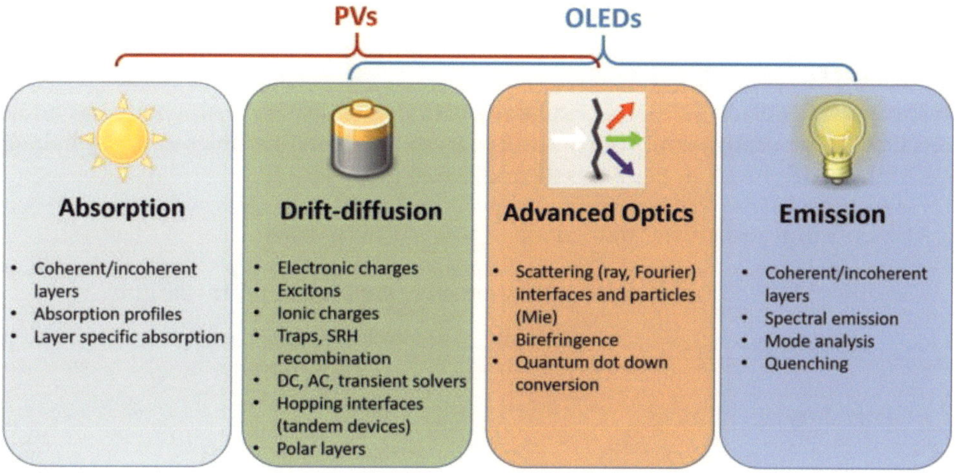

Figure 5.
Characteristics employed in SETFOS for PVs and OLEDS [24].

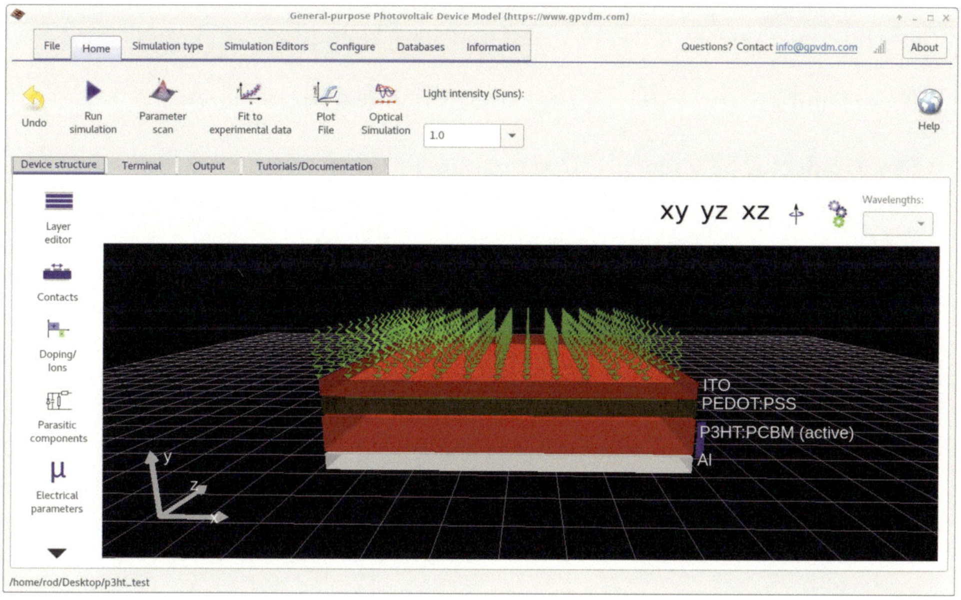

Figure 6.
Schematic representation of organic solar cell designed by GPVDM and user Interface in GPVDM.

6.8 AFORS-het

Among other simulators, AFROS-HET is one-dimensional numerical simulating software for modeling heterojunction PV solar cells, optoelectronic devices as well as some communal solar cell characterization techniques [26]. This simulating software is generated by A. Froitzheim et al. of Hahn-Meitner-Institute Berlin, Germany in 2003 [26]. It is useful for the maximum attainable efficiencies as well as developing the designing mechanism for solar cells like crystalline and amorphous solar cells [27]. The Auger recombination, intra-band model and Hurkx model are incorporated by the latest version of AFROS-HET [28]. This unique simulator has been generally utilized for (multi-junction) thin-film solar cells well as (heterojunction) multi-layer solar cells. The typical input panels in AFORS-HET are illustrated in **Figure 7**. The latest version 2.4 of AFORS-HET explains the one-dimensional semiconductor equations like Poisson's equation, continuity equation for electrons/holes with the help of predictable differences under altered conditions, i.e.,

- Equilibrium and steady state mode.
- Steady state mode with small supplementary sinusoidal perturbations.
- Simple transient mode.
- General transient mode.

A group of different physical models have been applied. The class of electron and hole pairs may be designated either by Lambert–Beer absorption containing irregular surfaces and by using measureable transmission and reflection files also by measuring the plain surface internal reflections with the help of complex indices of reflection for the different layers.

Basic input parameter of AFORS-HET and associated physical models:

- Super band gap generation optical models (Optical parameters).
- Semiconductor bulk models (Layer parameters).

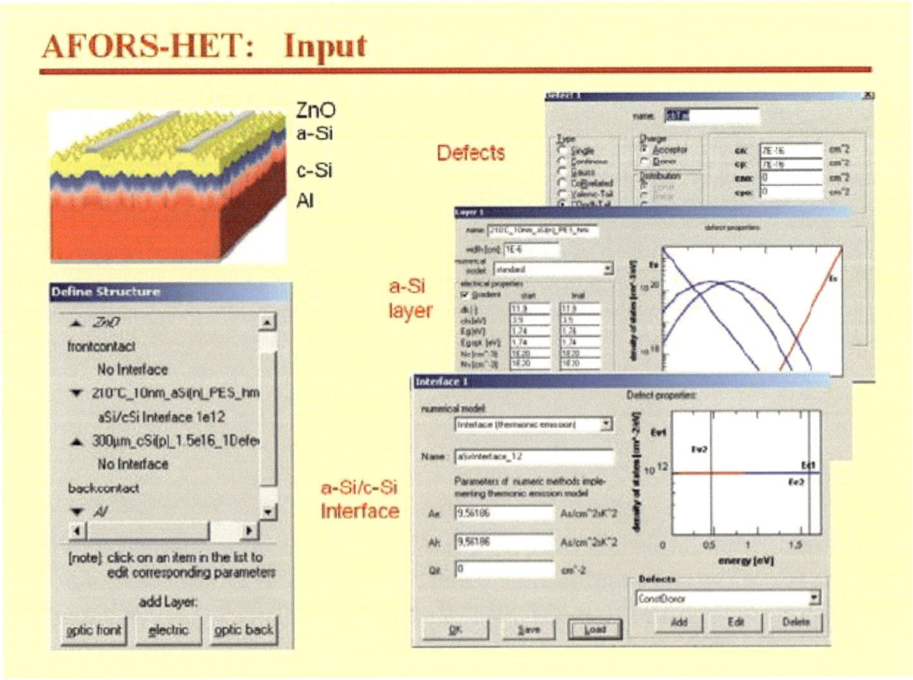

Figure 7.
Typical input panels in AFORS-HET.

- Semiconductor interface models (Interface parameters).
- Back/front contact to semiconductor boundary models (Boundary parameters).
- Circuit elements and external parameters.

6.9 Aspin-2D

ASPIN-2D is a particular two-dimensional computer program developed by M Vukadinović from Ljubljana University in 2000 [29]. This modeling tool is developed with the help of drift-diffusion model and has been utilized to simulate transport of heterojunction solar cells. The latest version of this software is ASPIN3 which permits the simulating lateral transport as well as mixtures of various materials, grain boundaries that are not promising with one-dimensional simulator [30]. This simulating package is suitable for p-i-n junction-based a-Si PV solar cell together with tandem cell and CIGS cell.

6.10 Photo electro chemical simulation software (PECSIM)

It is a state-of-the art programming platform for solar cell, designed by Matthias Schmid et al. at Zürich University in 2011 [31]. This simulator was specifically established for DSSCs (Dye-sensitized solar cells) using a certified optical model which includes an electrical model and ray-tracing algorithm. A unique and user-friendly GUI is unified in PECSIM for enabling the consumers to distinguish a deep vision into the contact between the various constituents of a DSSCs or perovskite solar cells (PSCs) as well as GUI supports the consumers to evaluate the conversion losses and embrace the optimization techniques for getting better the performance of solar cell. With the growing reputation of PSCs and DSSCs, the uses and applications of PECSIM simulating software have been growing extraordinarily.

6.11 Adept

ADEPT is another commonly used simulator developed by Jeffrey L. Gray in 1991 and written in C++ language [32]. This modeling software utilizes a comprehensive Newton method for execution by using ADEPT tool. One-dimensional simulation can be executed simply in usual PC environment. A latest uniting of the sparse matrix solvers permits this tool to accomplish two-dimensional simulations as well as it can be used to arrange the homo and heterostructures of PV solar cell and for further applications. ADEPT solves Poisson's equation and the hole and electron continuity equations in one dimension in compositionally non-uniform semiconductors. It was designed to represent solar cells made from amorphous silicon, copper indium diselenide and cadmium telluride. As the user enter material properties (band gap, mobility, etc.), devices made from any material with known parameters can be represented. At any operational point, carrier density, recombination, electric field and others may be displayed. The novel Frozen Potential Method simulates non-superposition solar cells. This simulator has been utilized to simulate the workings of GaAs, CIGS, AlGaAs, CdTe as well as a-Si:H PV solar cell [33].

6.12 TCAD

TCAD is a particular computer program generally known as commercial simulation package that has been developed for designing and fabrication of semiconductor device.

This solar simulator is designed by Synopsys that enable the programmers, researchers as well as industrialists to implement the device simulations for composite structured PV solar cell along with investigating the obtained simulation results [34, 35].

Many research organizations like NREL are using this solar simulator for calculating the CdTe, CIGS, GaAs as well as multi-junction PV solar cell efficiency and actual manufacturing of these solar cells. Most of the solar softwares revealed above are generated for one-dimensional solar cell simulations [36].

6.13 Atlas

ATLAS is an excellent solar simulator that has been generated for estimating the optoelectronic performance of 2D/3D semiconductor devices. This solar simulator is established by Silvaco Atlas and is utilized for simulation of various kinds of single junction PV solar cells. ATLAS has supplementary feature of simulation for tandem cell productions and performances. This simulating tool is also proficient to simulate the optoelectronic properties of inorganic and organic cells. Yet, this solar simulator has been performed for CIGS cell, textured cell, MIS cell, 3D coaxial cell and image sensors working simulation. The above discussed solar cell softwares are well reputable and commonly operated for executing the simulation task of multiple types of PV solar cells [37].

7. Other solar cell simulators

7.1 Open solar

It is the world's first open-access, end-to-end solar design and sales application that provides solar experts with a sophisticated yet user-friendly software package that meets all of their needs, from sales and lead management to solar system design, installation and analysis [38]. It comprises a database including average worldwide daily solar radiation statistics taken on the ground for each month of the year. The peculiarity of this solar database is that it is available to everyone, open for data extraction and data introduction. Data can be extracted by anybody. This extraction may be accomplished in a variety of methods, including downloading, online copy/paste, connecting to a web service or using Flex Remote Object Services (amf format).

Key Features:

Three-dimensional Design, Prominent Precision: The fastest, easiest and most accurate 3D design tool available. You can create your designs with reliability and affordability in the lab in in the various field.

Unified Business Partners: Easy selection and combination of your selected finance suppliers, real-time agreements, creating selling at ease than ever.

Tenders that Sell: Interactive proposals, fully customizable online having 24% sale adaptation strained from knowledge selling tens of thousands of systems live and over the cell phone.

Open Access: It is free for all users for an indefinite period. Link your current CRM and corporate tools to this production-leading platform.

7.2 PVsyst

This unique modeling software is introduced here for studying, analyzing data and designing complete PV systems. It includes multiple parameters for grid-connected, pumping, stand-alone and DC-grid PV systems. In addition, extensive databases of photovoltaic system component, as well as a range of broad-spectrum solar energy utensils are provided. PVsyst is intended for architects, engineers and researchers. It includes a user-friendly project development guide as well as a comprehensive contextual help menu that explains the approaches and models employed. PVsyst receives meteorological and personal data from a variety of sources. PVsyst produces a complete report, graphs and tables, as well as exportable data. This simulator is generally developed for solar designers, researchers and engineers. It is enormously favorable for learning and training [39]. The PVsyst system's design board is demonstrated in **Figure 8**.

Key Features:

- System sizing and designing.
- It generates simulations and outcomes.
- Model storage systems.
- Simulates the aging effect of solar modules, etc.
- Creating a shading scene.
- Introducing data and components and receiving climate data from metronome.

7.3 Solar labs

Solar Labs are unique and effective software for solar designing as well as sales growth in photovoltaic community. It concentrates in creating software for PV solar installers and commercial sectors to develop pilot sales quotes and elevate system proposals. They were established in 2017 and its headquarters is in India [40].

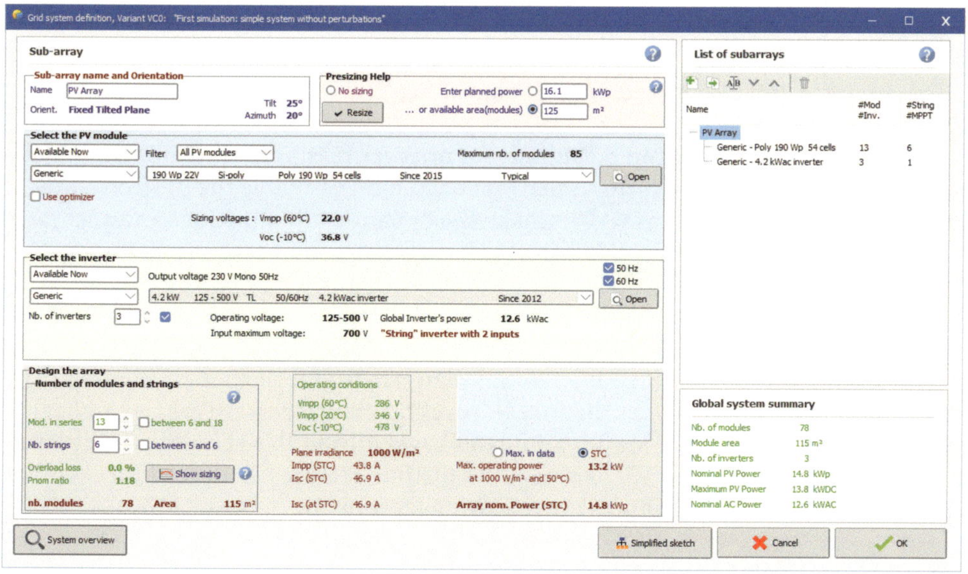

Figure 8.
System design board of PVsyst.

Design Formation: This simulator generates designs that are simple, easy, accurate, reliable and fast, which can be viewed in 2D or 3D. Also creates various rehearsals and picks the best reliable solution liable on customer's requirements.

Smarter Energy Modeling and Profiling: It develops thorough electrical figures and diagrams for domestic and commercial projects in a few minutes. You can easily illustrate the mounting templates and panel tilt values by selecting from a range of options. Draw-in barriers of any figure with the help of the polygon tool and others.

Precise Simulations and Shading Analysis: Without difficulty calculate your monthly, annual power consumption and production. Generate shading patterns of any single day of the year and simulate reliable heat maps according to your proposals.

Trade Proposals and Profit-making Reports: Invite clients with systematic and far-reaching attractive proposals. Attention on the basic factors like vital growing metrics and make available several financial options in a brief mode. Observe your aggregate savings, Return Period and Interior Return Rate in one single assessment.

7.4 Aurora solar

Aurora Solar presents unique software that allows solar photovoltaic engineering design as well as gives planned management performance, and gives reliable facilities for sales and customer services that may be helpful for solar PV installers as well as financiers. Aurora Solar software is found by Aurora Christopher Hopper and Samuel Adeyemo in 2013, headquarters situated in San Francisco, California, USA [41].

Key Features:

- Efficient and Simple Design Interface:

Users can utilize comprehensive shade measurement and shading analysis rightly in solar PV applications, effective components for system designs and fabrication also to create ID and 2D diagrams automatically. Through this simulator all generated designs are checked automatically for electrical constraints, NEC rule compliance and industrial interpretations and applications.

- Automated and easily accessible Financial Report:

This simulator provides rapid findings for project's feasibility, streamline models and most reliable financing formations. Financial models enable the project's competency for accurate profile as well as make the dynamic visualizations and accuracy reliable of financial information like payback periods, cash flows and bill savings.

7.5 HelioScope

In solar industry, HelioScope is a famous platform for fabrication, designing high-performance solar panels. It has distinct features that are necessary for researcher's interest in solar community. Many candidates have utilized this simulating software to fabricate and design solar PV arrays. It has distinct features as an easily accessible solar design tool. This contributes the following features given below [42–44].

- PV Solar Panels Layout.

- Indefinite Designs.

- CAD (Computer-Aided Drawing) Tools.
- Google Charts Integration.
- Energy Loss Map.
- Single-Click Sharing.
- Quick Design Revision.
- Calculation for voltage Drop.
- Wiring Variety.
- PAN File Backing.
- Drawing of Shading Integration.
- Single Line Diagram Export.
- NSRDB or NREL Meteo Integration.
- Estimates of Solar Energy Production.

8. Results and discussion

Now a time, there are number of photovoltaic modeling softwares reported in literature covering the areas of semiconductor physics, photonics, optics, electrical circuit design also in device fabrication and characterization along with costs analysis [5, 44]. This chapter describes a well-established solar simulators like PC1D, SCAPS-1D, wxAMPS-1D, AMPS-1D, ASA, Gpvdm, SETFOS, PECSIM, ASPIN, ADEPT, AFORS-HET, TCAD and SILVACO ALTAS.

At the beginning, many simulating softwares have their parameters, variables and function set up in a variety of scenarios, light intensities and device properties. In order for photovoltaic cells to function properly, light must first be taken in and then used to create an electron–hole pair. At the first stage of this process, the energy contained in a photon is transformed into the form of electrical energy by means of the formation of an electron–hole pair. Because of the natural variance that occurs in the terrestrial spectrum, every reference spectrum has a degree of arbitrary quality to it. The AM1.5G global spectrum is used in order to ensure that the standard spectrum is reflective of the vast majority of PV cells. It is a common practice to assume that the power density is 1000 W/cm^2. The photon flux of the AM1.5G spectrum will be utilized to determine the generation rate. The temperature of the ambient environment is assumed to be 300 K unless specified otherwise. The semiconductor module will provide the semiconductor function such as doping, generation and recombination, trap density and space charge density, based on Poisson's Equation, in order to analyze the carrier transport with the gradients of semiconductor parameters around the heterojunction interfaces.

PC1D is suitable simulator for crystalline silicon solar cell whereas SCAPS-1D is for CIGS, ASA is for amorphous silicon, SETFOS, PECSIM and Gpvdm are employed for organic solar cell and AFROS-HET is for heterojunction solar cell, AMPS-1D, SCAPS-1D, ADAPT and TCAD are used for multiple solar cell simulations. Most of the simulating softwares have been generated by using a single junction solar cell model. ASA, SETFOS and ADEPT have additional features for simulating the comparatively lower efficiency tandem solar cells. But wxAMPS and TCAD simulators

S/N	Simulators	Dimension	Availability	System requirement	Capability
1	PC1D	1D	Free of cost	All windows versions up to Windows 7	Capable to simulate most of the solar cell but perfect for (C-Si) solar cell.
2	ASA	1D	Paid	Windows 7 or 8 operating systems	Thin Si:H/μc -film ce -Si:H tandem cell ll: a-Si:H, μc-Si:H and a-.
3	AMPS	1D	Free of cost	MacOS 10.9 and windows 7 or upgrade	C-Si, a-Si:H, CIGS, CZTS cells.
4	wxAMPS	1D	Free of cost	All windows versions up to Windows 7, Vista	Having all feature of AMPS including tandem solar cell.
5	SCAPS	1D	Free of cost	Compatible with Windows and Linux	Thin-film Solar cell: CIGS, CdTe, GaAs, C Si and a-Si:H cell.
6	SETFOS	1D	Paid	Compatible Windows and Linux (x86 and ARM)	Organic, perovskites, quantum-dots, perovskite/silicon tandem solar cells.
7	Gpvdm	1D	Paid	Windows Vista/7/8 systems	perovskite, polymer, C-Si, a-Si:H and CIGS cells.
8	AFORS-HET	1D	Free of cost	Windows XP / Vista / 7 / 8 / 8.1/Linux	a-Si:H/c-Si cell.
9	ASPIN	2D	Free of cost	Windows 7 / 8.1 /10	CIGS, a-Si:H/C-Si. a-Si:H p-i-n heterojunction cell.
10	PECSIM	N/A	Free of cost	Windows 7 / 8.1 /10	Perovskite, dye-sensitized solar cells (DSCs).
11	ADEPT	1D	Free of cost	Windows 10, Windows Server 2016, Processor: Intel Core i7 or equivalen	C-Si, GaAs, AlGaAs/GaAs tandem, CIGS, CdTe and thin-film a-Si:H solar cell.
12	TCAD	2D, 3D	Paid	Windows 7, 8, 8.1 and 10 (64-bit), Linux 6 & 7	CMOS, power, memory, image sensors, solar cells.
13	ATLAS	2D, 3D	Paid	Windows and Linux platforms (version not mentioned)	Organic solar cell, tandem solar cell, photodetector.

Table 1.
Comparative study of solar cell simulators.

have III-V multi-junction solar cell simulating capabilities as compared to single junction solar cell.

The research indicates that 3D models offer more accurate estimates of the parasitic losses that occur in solar cells. But nonetheless, one more method for calculating these parasitic losses is to use the equivalent electrical circuit of the solar cell and the IV response equation. In this particular scenario, the 1D simulation will serve as the foundation for defining PV devices in such a way that they are able to adequately manage the complicated carrier movement.

It is important to note that the majority of PV simulators provide simulation capabilities at no cost, with the exception of paid programmes such as ASA, SETFOS, SILVACO ATLAS and TCAD. This fact has to be brought to your attention. The availability of these tools, along with their compatibility with various applications and operating systems, is outlined in **Table 1**. This comparative research sheds light on the precise instrument that should be used to simulate solar cells.

9. Conclusion

In this article, we gave an overview of the various open-source software tools that are available for modeling photovoltaic (PV) cells used in wearable applications. In total, we reviewed eighteen different software programmes. According to the findings of our investigation, the various simulation technologies delivered a variety of outcomes. This was owing to the fact that they used different approaches to the problem of solving the semiconductor equations. In spite of this, we came to the conclusion that PC1D and SCAPS-1D is the best accurate non-commercial tool for use in wearable applications. It has a straightforward operation technique, a friendly dialog box, extremely quick simulations at no additional expense and support for multi-junction solar cells. This study also includes a list of helpful references and internet connections for the photovoltaic community. These references and sites might be of use when choosing the best simulators for acceptable solar cells. This chapter may be useful for early career researchers, professional programmers and seasoned researchers in the field of solar cell simulation in the process of constructing a variety of various innovative solar cell simulators.

Additional information

ORCID ID: 0000-0003-4485-3315.

Author details

Abdul Shakoor[1*], Ghazi Aman Nowsherwan[2], Muhammad Fasih Aamir[3], Ahmar Ali[4], Sami Ur Rehman[5], Waheed Alam[5], Muhammad Yasir[6], Khizra Arif[6], Muhammad Ahmad[2] and Jamal Yousaf[7]

1 Department of Physics, University of the Punjab, Lahore, Pakistan

2 Centre of Excellence in Solid State Physics, University of the Punjab, Lahore, Pakistan

3 International Center for Materials Nanoarchitectonics (WPI-MANA), National Institute for Materials Science (NIMS), Tsukuba, Japan

4 College of Engineering and Physics, Physics Department, King Fahd University of Petroleum and Minerals (KFUPM), Dhahran, Saudi Arabia

5 Department of Physics, Ripah International University, Islamabad, Pakistan

6 Department of Material Science and Engineering, Insitute of space technology, Islamabad, Pakistan

7 Department of Physics, University of Wah, Wah Cantt, Pakistan

*Address all correspondence to: abdulphysics2020@gmail.com

References

[1] Basore PA, Rover DT, Smith AW. PC-1D version 2: Enhanced numerical solar cell modelling. In: Record of the Twentieth IEEE Photovoltaic Specialists Conference. USA: IEEE; 28 Sept 1996:389-396

[2] Rover DT, Basore PA, Thorson GM. Solar cell modeling on personal computers. In: IEEE Photovoltaic Specialists Conference. USA: IEEE; Vol. 18. 1985

[3] Liu Y, Ahmadpour M, Adam J, Kjelstrup-Hansen J, Rubahn H-G, Madsen M. Modeling multijunction solar cells by nonlocal tunneling and subcell analysis. IEEE Journal of Photovoltaics. 2018;**8**:1363-1369

[4] Haddout A, Raidou A, Fahoume M. A review on the numerical modeling of CdS/CZTS-based solar cells. Applied Physics A. 2019;**125**:1-16

[5] Sharma DK, Verma V, Singh AP. Review and analysis of solar photovoltaic softwares. International Journal of Current Engineering and Technology. 2014;**4**:725-731

[6] Lalwani DKAMSM. Investigation of solar photovoltaic simulation softwares. International Journal of Applied Engineering Research. 2010;**1**:581-601

[7] Haug H, Greulich J, Kimmerle A, Marstein ES. PC1Dmod 6.1–state-of-the-art models in a well-known interface for improved simulation of Si solar cells. Solar Energy Materials and Solar Cells. 2015;**142**:47-53

[8] Hashmi G, Akand AR, Hoq M, Rahman H. Study of the enhancement of the efficiency of the monocrystalline silicon solar cell y optimizing effective parameters using PC1D simulation. Silicon. 2018;**10**:1653-1660

[9] Hashmi G, Rashid MJ, Mahmood ZH, et al. Investigation of the impact of different ARC layers using PC1D simulation: Application to crystalline silicon solar cells. Journal of Theoretical and Applied Physics. 2018;**12**:327-334

[10] Zeman M, Willemen JA, Vosteen LLA, Tao G, Metselaar JW. Computer modelling of current matching in a-Si: H/a-Si: H tandem solar cells on textured TCO substrates. Solar Energy Materials and Solar Cells. 1997;**46**:81-99

[11] Isabella O, Solntsev S, Caratelli D, Zeman M. 3-D optical modeling of thin-film silicon solar cells on diffraction gratings. Progress in Photovoltaics: Research and Applications. 2013;**21**:94-108

[12] Hernández-Como N, Morales-Acevedo A. Simulation of heterojunction silicon solar cells with AMPS-1D. Solar Energy Materials and Solar Cells. 2010;**94**:62-67

[13] Mostefaoui ABAR. AMPS-1D in a-Si: H/µc-Si: H tandem solar cells. Journal of Applied Sciences. 2011;**11**:2932

[14] Bouloufa A, Djessas K, Zegadi A. Numerical simulation of CuInxGa1–xSe2 solar cells by AMPS-1D. Thin Solid Films. 2007;**515**:6286-6287

[15] Zhu H, Kalkan AK, Hou J, Fonash SJ. Applications of AMPS-1D for solar cell simulation. AIP Conference Proceedings. 1999;**462**:309-314

[16] Liu Y, Sun Y, Rockett A. A new simulation software of solar cells—wxAMPS. Solar Energy Materials and Solar Cells. 2012;**98**:124-128

[17] Yaşar S, Kahraman S, Çetinkaya S, Apaydın Ş, Bilican İ, Uluer İ. Numerical

thickness optimization study of CIGS based solar cells with wxAMPS. Optik. 2016;**127**:8827-8835

[18] Çetinkaya S. Study of electrical effect of transition-metal dichalcogenide-MoS2 layer on the performance characteristic of Cu2ZnSnS4 based solar cells using wxAMPS. Optik. 2019;**181**:627-638

[19] He J, Zhang T, Tang P, Qiu C, Feng X, Luo X. Simulation on the carrier transport process inside the semiconductor of dye sensitized solar cells by wxAMPS software. Electrochimica Acta. 2014;**125**:218-224

[20] Olopade MA, Oyebola OO, Adewoyin AD, Emi-Johnson DO. Modeling and simulation of CZTS/CTS tandem solar cell using wxAMPS software. In: IEEE 42nd Photovoltaic Specialist Conference (PVSC). USA: IEEE; Jun 2015:1-4

[21] Burgelman M, Nollet P, Degrave S. Modelling polycrystalline semiconductor solar cells. Thin Solid Films. 2000;**361**:527-532

[22] Burgelman M, Decock K, Khelifi S, Abass A. Advanced electrical simulation of thin film solar cells thin solid films. Thin Solid Films. 2013;**535**:296-301

[23] Neukom MT, Schiller A, Züfle S, Knapp E, Ávila J, Pérez-del-Rey D, et al. Consistent device simulation model describing perovskite solar cells in steady-state transient and frequency domain. ACS applied materials & interfaces. 2019;**11**(26):23320-23328

[24] Setfos-Introduction. 2019. Available from: https://www.fluxim.com/setfos-intro [Accessed: Sepember 9, 2019]

[25] MacKenzie RCI, Kirchartz T, Dibb GFA, Nelson J. Modeling nongeminate recombination in P3HT: PCBM solar cells. The Journal of Physical Chemistry C. 2011;**115**:9806-9813

[26] Froitzheim A, Stangl R, Elstner L, Kriegel M, Fuhs W. AFORS-HET: A computer-program for the simulation of heterojunction solar cells to be distributed for public use. In: 3rd World Conference onPhotovoltaic Energy Conversion. USA: IEEE; May 2003;**1**:279-282

[27] Stangl R, Kriegel M, Schmidt M. AFORS-HET, version 2.2, a numerical computer program for simulation of heterojunction solar cells and measurements. In: IEEE 4th World Conference on Photovoltaic Energy Conference. USA: IEEE; May 2006;**2**:1350-1353

[28] AFORS-HET: Version 2.5. [Online]. Available from: https://www.helmholtzberlin.de/forschung/oe/ee/si-pv/projekte/asicsi/aforshet/changelog_en.html [Accessed: Sepember, 2019]

[29] Vukadinović M, Smole F, Topič M, Krč J, Furlan J. Numerical modelling of trap-assisted tunnelling mechanism in a-Si: H and μc-Si n/p structures and tandem solar cells. Solar Energy Materials and Solar Cells. 2001;**66**:361-367

[30] ASPIN3. [Online]. Available from: http://lpvo.fe.uni-lj.si/en/software/aspin3/ [Accessed: Sepember 9, 2019]

[31] Wenger S, Schmid M, Rothenberger G, Gentsch A, Grätzel M, Schumacher JO. Coupled optical and electronic modeling of dyesensitized solar cells for steady-state parameter extraction. The Journal of Physical Chemistry C. 2011;**115**:10218-10229

[32] Gray JL. Adept: A general purpose numerical device simulator for modeling solar cells in one-, two-, and three-dimensions. In: The Conference Record of the Twenty-Second IEEE Photovoltaic Specialists Conference-1991. USA: IEEE; 7 Oct 1991. pp. 436-438